Weed Science: Principles and Applications

Weed Science: Principles and Applications

Remi Ford

MURPHY & MOORE
www.murphy-moorepublishing.com

Published by Murphy & Moore Publishing,
1 Rockefeller Plaza,
New York City, NY 10020, USA
www.murphy-moorepublishing.com

Weed Science: Principles and Applications
Remi Ford

International Standard Book Number: 978-1-63987-568-9 (Hardback)

Cataloging-in-Publication Data

Weed science : principles and applications / Remi Ford.
p. cm.
Includes bibliographical references and index.
ISBN 978-1-63987-568-9
1. Weeds. 2. Weeds--Control. 3. Herbicides. I. Ford, Remi.

SB611 .W44 2022
632.5--dc23

TABLE OF CONTENTS

PREFACE

The plants which are considered undesirable or troublesome and grow in places where they are not desired are known as weeds. Some common examples of everyday weeds are thistle, garlic mustard and dandelion. Weed science is defined as the study of effects of weed upon human activities and their management in agricultural production systems, natural areas, and managed properties in residential and urban areas. It is a branch of applied ecology which seeks to modify the environment against natural evolutionary trends. Methods such as the use of herbicides, wilting with high heat, powered cultivation with cultivators, cultivation with hoes and burning are often used to control the growth of weed. The topics included in this book on weed science are of utmost significance and bound to provide incredible insights to readers. While understanding the long-term perspectives of the topics, it makes an effort in highlighting their impact as a modern tool for the growth of the discipline. Those in search of information to further their knowledge will be greatly assisted by this book.

To facilitate a deeper understanding of the contents of this book a short introduction of every chapter is written below:

Chapter 1- Weed science is a branch of applied ecology which deals with the weeds, their management and their effects on human activities. Weeds are the undesirable and wild plants that compete with the cultivated plants. This is an introductory chapter which will introduce briefly all the significant aspects of weed science.

Chapter 2- There are various criteria on the basis of which weeds can be classified such as anatomy, cotyledenous character, nature of stem and association. On the basis of anatomy, weeds can be grouped into annual, biennial and perennial weeds. This chapter has been carefully written to provide an easy understanding of the classification of weeds.

Chapter 3- The plants which are generally not cultivated and are wild but have some companion plant effect, contribute to soil health, are edible, or add ornamental value are the beneficial weeds. There are many plant species which are beneficial like Aegopodium podagraria, Allium vineale, Capsella bursa-pastoris, etc. This chapter discusses in detail these types of beneficial weeds.

Chapter 4- The study of the establishment, growth, reproduction and life cycles of weeds is known as weed biology. The various stages related to weeds like weed seed production, weed seed dispersal, weed seed germination and various aspects of weed ecology like weed seed longevity, weed crop competition and weed thresholds, etc. are discussed in an easy-to-understand manner in this chapter.

Chapter 5- Weed control is concerned with the attempts to control weeds which compete with the desired flora and fauna. Soil steam sterilization is one method which is used to sterilize the soil with steam and to remove many pests such as weeds, bacteria, fungi, etc. The various methods of weed control like aquatic weed harvester, soil steam sterilization, soil solarization, stale seed bed and bioherbicide, etc. are closely examined in this chapter.

Finally, I would like to thank the entire team involved in the inception of this book for their valuable time and contribution. This book would not have been possible without their efforts. I would also like to thank my friends and family for their constant support.

Remi Ford

Chapter 1

Introduction to Weed Science

Weed science is a branch of applied ecology which deals with the weeds, their management and their effects on human activities. Weeds are the undesirable and wild plants that compete with the cultivated plants. This is an introductory chapter which will introduce briefly all the significant aspects of weed science.

Weed

There are numerous definitions of a weed, including:

- A plant out of place and not intentionally sown.
- A plant growing where it is not wanted.
- A plant whose virtues have not yet been discovered.
- Plants that are competitive, persistent, pernicious, and interfere negatively with human activity.
- And many others.

No matter what definition is used, weeds are plants whose undesirable qualities outweigh their good points, according to man. Our human activities create weed problems since no plant is a "weed" in nature. Though we may try to manipulate nature for our own good, nature is persistent. Through the manipulation process, certain weeds are controlled, while, other more serious weeds may thrive because favorable growing conditions for them also have been meet. Weeds are naturally strong competitors and those weeds that can best compete always tend to dominate. Both humans and nature are involved in plant breeding programs. The main difference between the two programs is that man breeds plants for yield, while nature breeds plants for survival.

Characteristics of Weeds

Certain characteristics are associated with and allow the survival of weeds. Weeds posses one or more of the following:

- Abundant seed production.
- Rapid population establishment.
- Seed dormancy.
- Long-term survival of buried seed.

- Adaptation for spread.
- Presence of vegetative reproductive structures.
- Ability to occupy sites disturbed by human activities.

There are approximately 250,000 species of plants worldwide; of those, about 3% or 8000 species behave as weeds. Weeds are troublesome in many ways. Primarily, they reduce crop yield by competing for water, light, soil nutrients, and space. Other problems associated with weeds in agriculture include:

- Reduced crop quality by contaminating the commodity.
- Interference with harvest.
- Serve as hosts for crop diseases or provide shelter for insects to overwinter.
- Limit the choice of crop rotation sequences and cultural practices.
- Production of chemical substances which are toxic to crop plants (allelopathy), animals, or humans.

Costs of Weeds

Weeds are common on all 485 million acres of U.S. cropland and almost one billion acres of range and pasture. Since weeds are so common, people generally do not understand their economic impact on crop losses and control costs. In 1991, the estimated average annual monetary loss caused by weeds with current control strategies in the 46 crops grown in the United States was $4.1 billion. If herbicides were not used, this loss was estimated to be $19.6 billion. Losses in field crops accounted for 82% of this total.

Another source estimates that U.S. farmers annually spend $3.6 billion on chemical weed control and $2.6 billion for cultural and other methods of control. The total cost of weeds in the United States could approach $15 to $20 billion dollars (Ashton and Monaco, 1991). Also, weed control and other input costs (e.g., seed, fertilizer, other pesticides, fuel) vary with the crop. For example, in the mid-90s, herbicides for soybeans cost $30/acre or about 47% of the $63/acre in total purchased input. For corn, the cost was $32/acre or about 28% of the $114/acre in total purchased input. And for wheat it was $6 or about 6% of the total $96/acre inputs. Several factors help determine the relative costs of herbicides from one crop to another and include the competitive ability of the crop, the weeds present, the contribution of non-chemical control practices, the tillage method, management decisions, and the value of the crop.

Benefits of Weeds

Despite the negative impacts of weeds, some plants usually thought of as weeds may actually provide some benefits. Some attributes include:

- Soil stabilization.
- Habitat and feed for wildlife.

- Nectar for bees.
- Aesthetic qualities.
- Add organic matter.
- Provide genetic reservoir.
- Human consumption.
- Provide employment opportunities.

Weeds have a controversial nature. But to the agriculturist, they are plants that need to be controlled, in an economical and practical way, in order to produce food, feed, and fiber for humans and animals. In this context, the negative impacts of weeds indirectly affect all living beings.

An undeniable and expensive consequence of agricultural practices is the adaptation of weeds to agricultural systems. Weeds are responsible for significant crop yield losses and for financial losses in agricultural production – in the order of 10% per year worldwide. To address this critical problem, the discipline of weed science has expanded over the past 50yr into an amalgam of scientists and practitioners who ask a variety of questions and employ a myriad of tools focused on understanding and managing weeds. Two predominant foci of this discipline are the fundamental aspects of weed biology and ecology, and the practical aspects of managing these pests. While weeds can infest many types of ecosystems, weed scientists focus on how weed populations affect crop yield and how to best apply this knowledge to prevent, eradicate or control weeds, primarily through the use of herbicides, with the ultimate goal of maximizing crop production.

Evolutionary biologists study various aspects of plant biology in order to gain a better understanding of evolutionary processes. Unfortunately, evolution has not been a significant component of weed science, and the question 'What makes these species inherently weedy?' often gets lost between the two fields of study. However, just as crops have been domesticated to serve the needs of human agriculturalists, weeds have been domesticated indirectly by humans for adaptation to agricultural fields. All weed-control methods – including nonchemical methods – exert selection pressure on weed populations, which effectively favors weeds that are better adapted to the method in use.

An extreme example of weed selection is herbicide resistance in response to the repeated use, over several years, of the same herbicide or mode of action, which imposes heavy selection pressure favoring individuals in the population which can survive that herbicide. Resistance management requires the reduction of herbicide selection pressure through the use of alternative methods of weed control. It naturally follows that considering evolutionary principles when managing resistant weeds should reduce, or even prevent, the evolution of greater resistance problems. Furthermore, to gain a clearer picture of how plants become weeds, we need to understand how the genetic architecture underlying weed-related traits interacts with all management practices, not just herbicides, which are ostensibly the environment of the weed.

The Continuing Evolution of Weeds

Weeds, defined anthropocentrically as undesirable plants that are growing 'out of place', have evolved numerous mechanisms to survive field conditions that are optimized for crops. Certain traits are generally associated with weedy plant species, including early germination, rapid growth from seedling to sexual maturity, and the ability to reproduce sexually and asexually. The 'ideal weed' might exhibit a generalist genotype with a high level of phenotypic plasticity. However, weeds are known to evolve rapidly in several ways: from colonizers selected by agricultural practices; from hybridization between wild and domesticated cultivars; and from selection on abandoned domesticated cultivars. Thus, the question of whether weeds are phenotypically plastic, 'Jacks of all trades', or possess single genes/traits that are responsible for weediness, remains unresolved.

There are many well-known examples of the appearance of new weeds or weed complexes following selection by a weed-control practice or cropping system. Annual tillage systems are known to select annual weeds and to disfavour perennial weeds, while perennial and no-till cropping systems generally select perennial weeds. Mowing often selects weeds with horizontal growth form and low-growing meristems, such as grasses and clovers, or prostrate phenotypes within the same species. In some systems weeds have evolved that mimic crops morphologically and phenologically, such as barnyardgrass (Echinochloa crus-galli) growing in rice (Oryza sativa). There are also many examples of biological agents that have had profound effects on weed floras, such as the increase in noxious thistles and other unpalatable species in response to grazing animals on rangelands. In some cases the result is simply a shift or a change in the species composition of a site, similar to succession in natural systems, while in other cases rapid weed evolution has occurred. Weeds are 'potentially useful for studies of microevolution under human influence'.

Evolutionary Ecology in Agricultural Ecosystems

The very traits that are a bane to those attempting to manage weed infestations are also traits that make weeds excellent scientific model organisms. In fact, many agronomic weeds that have a significant, negative economic impact are also models in ecology and evolutionary biology. For example, morning glories (Ipomoea spp.), the wild radish (Raphanus), and weedy sunflower and thistles (various genera) are all well represented by some of the major research threads in evolutionary ecology, such as studies of plant mating systems, plant–herbivore interactions, ecological adaptation and speciation. Likewise, representative species from these groups are regularly listed among the top 10 'worst' weeds in current-day agriculture and as such are often the subject of applied studies. Understanding the mechanisms that promote weed persistence, including herbicide resistance, is of practical significance; it also provides us with a model for understanding the genetics of adaptation.

From an evolutionary perspective, the principles and practices of agriculture create a large experiment across the landscape. The integration of the practical side of weed science with hypothesis-driven evolutionary ecology, along with the tools of genetics and genomics, will provide answers to the questions of not only the types of mutations that arise to promote weed persistence and vigor, but also how many genes might be involved in these traits, how they might impact one another, and the specific ecological context that can promote or constrain their persistence in weed populations. Adding to the complexity of the evolutionary ecology of weeds, however, is their phylogenetic diversity, which probably precludes the use of a single model weed species.

Common Weeds

Poison Sumac (Toxicodendron Vernix)

Poison sumac is a shrub (some consider it a small tree) that grows in swampy areas, often next to Cinnamon ferns and cattails. You will not find it trailing over the ground or climbing trees, as you sometimes find poison ivy. Every part of the plant is poisonous, meaning it can cause serious rashes if touched. As is often the case with toxic plants, it can also be very attractive; its white berries and bright fall foliage is pretty as well as dangerous.

Japanese Knot Weed (Polygonum Cuspidatum)

Polygonum cuspidatum goes by several other common names, including Japanese knotweed and fleece flower. Several other common names include the term, “bamboo,” such as “Mexican bamboo.” While its autumn flower does, indeed, look fleecy, “fleece flower” is just too dainty a name for so tenacious a weed.

Crabgrass (Digitaria)

Being an annual weed, crabgrass perpetuates itself via seed—millions of seeds. To control crabgrass, you'll need to address the issue in spring when the plant is at its most vulnerable. The best option is to remove the plants by hand, roots and all. After that, use an organic fertilizer to encourage the growth of lawn grass which will crown the crabgrass out.

Dandelions (Leontodon Taraxacum)

Dandelions are a harbinger of spring. Their bright yellow flowers often poke up through lawns and appear between cracks in driveways and sidewalks. The seed heads of dandelions are probably better known than those of crabgrass, but dandelions are perennial, not annual weeds.

While dandelions have multiple medicinal uses and can be eaten in salads or used to make wine, many homeowners would prefer to eliminate them. Keeping dandelion seeds from germinating won't be enough to get rid of dandelions. It's possible to use herbicide to eliminate your dandelions, but the most effective and least harmful approach is to dig the flowers up from the roots.

Plantain Plants (Plantago Major)

A rather innocuous plant, common plantain can simply be mowed whenever you mow the lawn. Its relative, Plantago lanceolata is a similar weed, but with narrow leaves. Now a ubiquitous lawn weed in North America, broadleaf or "common" plantain was brought to the New World by colonists from Europe for its medicinal uses. Common plantain has many medicinal uses. Mashed, it can be used as a poultice for bee stings; the leaves can also be dried and made into a tea to treat diarrhea.

Common Ragweed (Ambrosia Artemisiifolia)

Common ragweed may be an important weed for you to identify, even if you don't care about keeping your yard weed-free for aesthetic reasons. If you're an allergy sufferer, you should be aware that common ragweed is a major source of hay fever.

Giant Ragweed (Ambrosia Trifida)

It's not for nothing that this plant is named, "giant ragweed." It can grow up to 15 feet tall, with thick roots and branches. Like its ragweed cousin (and unlike goldenrod), giant ragweed produces a great deal of pollen which causes serious allergies.

Hedge Bindweed (Convolvus Arvensis)

Hedge bindweed has a fairly attractive bloom, similar to that of the morning glory, which can be white or pink and have a pleasant fragrance. But this is no innocuous weed. If you let hedge bindweed get out of control, your yard will feel like Gulliver in Lilliput. There is a reason for that "bind" in "bindweed."

Ground Ivy (Glechoma Hederaceae)

Ground ivy, a common lawn weed, goes by a number of names. For instance, it is also called "gill," "gill-over-the-ground" and "creeping charlie." Although considered a weed, ground ivy has a pretty flower and, when you mow this weed, it gives off a pleasing aroma. Ground ivy is also used as a medicinal herb.

Purslane (Portulaca Olearacea)

Purslane is the edible weed, par excellence. Purslane contains five times the amount of essential omega-3 fatty acid that spinach has, and its stems are high in vitamin C. A succulent mat-forming plant, it has a crispy texture and interesting peppery flavor. It is often served raw in salads but can also be cooked as a side dish.

Stinging Nettle (Urtica Dioica)

The flowers of stinging nettle plants are inconspicuous. You'll pay plenty of attention to its barbs, however, if you're unfortunate enough to brush against stinging nettle. The discomfort these weeds can cause seems incongruous with the fact that stinging nettle is edible. But the young leaves of stinging nettle are, indeed, cooked and eaten by wild foods enthusiasts. Just be sure to pick at the right time and prepare properly to ensure safe consumption.

Curly Dock (Rumex Crispus)

Curly dock (also called "curled dock" or "yellow dock") is more than just distinctive, it's also useful: curly dock serves as a home remedy to treat stinging nettle burns--though it can be toxic to consume. You'll be able to identify curly dock by its greenish blossoms that cluster long thin flower stocks. After the flowers have dried and turned brown, they remain in place, making the plant easy to recognize. The flowers start out a much less distinctive light-greenish or reddish color. Blooming occurs in clusters in the form of multiple, long, skinny flower stalks at the top of the plant.

Wild Madder (Galium Mollugo)

Wild madder is, like sweet woodruff, in the Galium genus. Wild madder is also called "bedstraw." Apparently, people did actually once use this weed as a bedding material. Sweet woodruff is a creeping, mat-forming perennial that pretty clusters of white star-shaped flowers in spring and has very fragrant, lance-shaped dark-green leaves.

Clover Leaf (Trifolium)

While many consider clover a "weed," there's really nothing wrong with having a little clover mixed into your lawn. The Irish consider various tripartite clover leaves (such as the one in the photo here) to be "shamrocks." The tradition behind the shamrock is quite distinct from that behind four-leaf clovers.

Orange Jewelweed (Impatiens Capensis)

Like curly dock, orange jewelweed (or "jewel weed") can be used as a home remedy for poison ivy. The taxonomic name of orange jewelweed, Impatiens capensis, classifies it as a wild version of the colorful impatiens flowers sold so widely for shady annual beds.

Bittersweet (Celastrus)

There are three plants named, "bittersweet." American bittersweet is harmless, but Oriental bittersweet should be regarded as a weed since it can harm your trees. The third type of weed that goes by this name (bittersweet nightshade) is one of our most poisonous plants, despite being related to the tomato.

Horsetail Weed (Equisetum Arvense)

There's more than one kind of "horsetail." Equisetum arvense is a thoroughly weedy-looking plant that will spread out of control if given a chance, even in dry soil. Equisetum hyemale, by contrast, is a more useful horsetail plant to the landscaper. It is an architectural plant that can be employed as an accent around water features. If given moist soil, it, too, will spread, so consider potting it up for use around water features so that you'll have firm control over it.

Garden Weeds

These are weeds that are prohibited or controlled by law on a federal or state level. Noxious weeds are highly destructive and difficult to control by ordinary cultural practices. We have divided this list of weeds into two sections: 1) Noxious and 2) Other common weeds that compete with vegetables, fruits, and crops but may have their own beneficial uses.

Noxious Weeds

The noxious weeds (on federal and/or state level) on this list include field bindweed, quackgrass, Canada thistle, yellow nutsedge, and buckhorn plantain. There are other noxious weeds that aren't on this list that are also problematic, such as Johnsongrass.

Bindweed (Convolvulus Arvensis)

Field bindweed is a hardy perennial vine that has been given many names, including perennial or wild morning-glory, creeping jenny, sheepbine, cornbind, and bellbine.

Bindweed is not the same as the ornamental annual morning-glory (in the genus Ipomea) which has a larger (2-inch wide) and more showy flower that can be white to blue or purple; it also has a thicker stem that is sometimes hairy and heart-shaped leaves that are 1 ½ inches wide and 2 inches or more long. The two species are easy to distinguish from each other.

An invasive from Eurasia, field bindweed is one of the most persistent and difficult to control weeds. It spreads from an extensive rootstock and from seed. And its roots are found to depths of 14 feet. Lateral roots becoming a secondary vertical root. A single field bindweed plant can spread radially more than 10 feet in a growing season. This extensive underground network allows for overwintering without foliage, and it can persist for many years in the soil.

Bindweed sprouts in late spring and can be seen throughout the summer. Though the plant's flowers are attractive, field bindweed can become a big problem in warm weather, when they spread ruthlessly.

How to Control Bindweed

Unfortunately, tilling and cultivation seems to aide bindweed spread. Fragments of vertical roots and rhizomes as short as 2 inches can form new plants. Field bindweed also is very drought tolerant and once established is difficult to control even with herbicides.

The best control is, as with most weeds, is prevention or early intervention. Seedlings of field bindweed must be removed before they become perennial plants. However, this need to be done when they're young—about 3 to 4 weeks after germination. After that, perennial buds are formed, and successful control is much more difficult.

Bindweed can grow through many mulches so you need to use landscape fabrics such as polypropylene and polyester or mulches such as black plastic or cardboard but also ensure that the edges of the covering overlap so that the bindweed stems can't find their way into the light. If holes are made in the fabric or plastic for plants, bindweed will grow through these holes. A landscape fabric placed over soil then covered with bark or other plant-derived product (e.g., organic matter) or rock will likely keep field bindweed from emerging. It might take more than 3 years of light exclusion before the bindweed dies. Once landscape fabric or other mulch is removed, new bindweed plants might germinate from seed in the soil; be sure to monitor the site for new seedlings. All parts of the bindweed plant are poisonous.

Quackgrass (Elytrigia Repens)

Quackgrass is a creeping, persistent perennial grass that reproduces by seeds. Its long, jointed, straw-colored rhizomes form a heavy mat in soil, from which new shoots may also appear.

How to Control Quackgrass

Try to dig out this fast-growing grass as soon as you see it in your garden, being sure to dig up the entirety of the plant (including the roots). Dispose of in your waste bin rather than the compost pile, as it will likely continue to grow in the latter.

Canada Thistle (Cirsium Arvense)

Canada thistle is an aggressive, creeping perennial weed from Eurasia (despite its name). It infests crops, pastures, and non-crop areas like ditch banks and roadside. Canada thistle reduces forage consumption in pastures and rangeland because cattle typically will not graze near infestations.

This weed reproduces by seed and whitish, creeping rootstocks which send up new shoots every 8 to 12 inches. Plants 2 to 4 feet tall, It is a colony-forming weed, reproducing asexually from rhizomatous roots (any part of the root system may give rise to new plants) or sexually from wind-blown seed. The plant emerges from its roots in mid- to late spring and forms rosettes. Then, it will send up shoots every 8 to 12 inches. The plants will grow 2 to 4 feet tall. You may spots its purple flowers are produced in July and August.

How to Control Canada Thistle

Canada thistle is difficult to control because its extensive and deep root system allows it to recover from control attempts. Horizontal roots may extend 15 feet or more and vertical roots may grow 6 to 15 feet deep. Seeds may retain viability 4+ years in the soil. The first plants need to be destroyed by pulling or hoeing before they become securely rooted. Look for Canada thistle above ground in early spring.

If Canada thistle becomes rooted, the best control is to stress the plant and force it to use stored root nutrients. It's at its weakest during the flowering stage in summertime; this is a good time to begin cultivation and destroy the roots and rootstock. One season of cultivation followed by a season of growing competitive crops such as winter rye, will go a long way toward eradication.

An approved herbicide, applied for two years in an established in a thistle-infested area, is an effective control. Usually, a combination of techniques is needed. Consult with your cooperative extension office for an approved herbicide and suggested program.

Believe it or not, Canada thistle is in fact edible—with some preparation required, of course. After the spines are meticulously removed, the leaves can be prepared like spinach. The stems are the most prized part, though their bristled outsides must be peeled first. Be sure to wear gloves.

Nutsedge (Cyperus Spp.)

Nutsedges are perennial weeds that superficially resemble grasses, but they are thicker and stiffer and V-shaped. Their leaves are arranged in sets of three from their base instead of sets of two as you would find in grass leaves. They are among the most problematic weeds for vegetable crops and can greatly reduce harvest yields. Yellow nutsedge has light brown flowers and seeds, while purple nutsedge flowers have a reddish tinge and the seeds are dark brown or black.

How to Control Nutsedge

If you have nutsedge, it's often an indicated that your soil drainage is poor or waterlogged. However, once nutsedge is established, it's very difficult to control. The best approach is to prevent establishment of the weed in the first place.

Remove small plants before they develop tubers. Tubers are key to nutsedge survival. If you can limit production of tubers, you'll eventually control the nutsedge itself. Most herbicides aren't effective against tubers.

Also, eliminate the wet conditions that favor nutsedge growth. Use mulches in landscape beds. Landscape fabrics are the best mulch because the sharp leaves of nutsedge can pierce other mulches.

Dating back to ancient Egypt, yellow nutsedge has historically been harvested for its tubers, which have a sweet, nutty flavor. Purple nutsedge tubers are also edible, but have a less pleasant, bitter taste.

Buckhorn Plantain (Plantago Lanceolata)

(Also called buck plantain, English plantain, narrow-leaved plantain). Buckhorn plantain is a common perennial weed more common in pastures, meadows, and lawns. This narrow-leafed weed reproduces and spreads by seeds.

How to Control Buckhorn Plantain

Buckhorn plantain is low-growing which makes it difficult to remove by hand. This plant has a long taproot so it can become drought-tolerant and difficult to control. So, to remove this weed, be diligent about pulling up young plants and destroying it before the plants go to seed. Learn how to scout and recognize young plants to help prevent early introductions from becoming persistent problems.

The best control is also preventative: grow a lush stand so the surface of the soil is shaded and prevents new seeds from getting established. As a last resort, several herbicides are effective on buckhorn plantain. The best time to spray is in the fall (late October to early December). Speak to your local cooperative for approved products. Yes, this weed is edible, especially when the leaves are young and tender. Enjoy it raw, steamed, boiled, or sauteed.

Troublesome Weeds

The following weeds—though not considered noxious—can still present a problem when they show up unwanted in gardens. However, weeds like lambsquarters or dandelions may actually be sought after for their nutritional content or benefit to pollinators, respectively.

Purslane (Portulaca Oleracea)

Purslane is actually considered noxious in at least one U.S. state. However, this is not a widespread categorization. So, why is purslane, an edible succulent plant, considered so troublesome? The answer goes back to the definition of weeds: Purslane can produce over 2,000,000 seeds per plant. Purslane also can reproduce vegetatively through its leaves, making it especially tough to eradicate. Many a gardener hoed purslane one day only to see it growing at full strength the next.

Purslane is an annual, succulent-like weed that reproduces by tiny black seeds and stem fragments. This weed appears in late spring or early summer and likes warm weather, fertile soil and moist garden beds.

How to Control Purslane

The primary method of management for common purslane is prevention. In home landscapes and gardens, this weed is generally managed by hand-weeding. Pull out this weed as soon as you see it and destroy the plant; this weed can live in your soil for years.

Mulching is also helpful, especially in garden beds. To be effective, organic mulches should be at least 3 inches thick. Synthetic mulches (plastic or fabric mulch) which screen out light and provide a physical barrier to seedling development, also work well. Fabric mulches, which are porous and allow flow of water and air to roots, are preferred over plastics. Combinations of synthetic mulches with organic or rock mulches on top are commonly used in ornamental plantings. Young purslane is edible. It's a nutritional powerhouse and a great addition to a salad or stir-fry.

Crabgrass (Digitaria Spp.)

Crabgrass is a low-growing, summer annual that spreads by seed and from rootings of nodes that lie on the soil. Unmowed, it can grow to 2 feet tall.

This weed appears from mid-spring through summer when the ground is warm. It grows well under dry, hot conditions. As an annual, crabgrass dies at the end of each growing season, usually at the first frost in the fall, and it must produce new seeds every year.

How to Control Crabgrass

Fortunately, crabgrass is fairly easy to manage. Controlling crabgrass before it sets seed is important, because the seeds can remain viable for at least 3 years in soil. In the lawn, mowing regularly is often all you need to prevent them from flowering and producing seed. Most experts recommend that you mow your lawn to a height of 2 to 4 inches and that you mow frequently enough to keep it within that range.

Also, if you keep a lawn, be sure to select grass adapted to your location so it's a healthy, thick lawn. Because seedling crabgrass isn't very competitive, a vigorously growing turf will crowd out new seedlings. Perennial ryegrass is the best competition for crabgrass. It also provides some insect control, as it emits a natural poison that gives some small, damaging bugs the "flu." Fertilizing is key and must be done in the spring and in the fall. Crabgrass thrives in compacted lawns, so aeration can help. A mixture of 1 pint of hydrogen peroxide, diluted to 3 percent, per 100 square feet of lawn can help eradicate the pesky plant.

In gardens, you easily can control crabgrass by mulching, hoeing, and hand pulling when the plants are young and before they set seed. You also can control this weed with solarization. Several chemical herbicides are available but often aren't necessary. Mulching with wood products (e.g. wood chips or nuggets), composted yard waste, or synthetic landscape fabrics covered with mulch will reduce crabgrass in shrub beds and bedding plants and around trees by blocking sunlight needed for its germination, establishment, and growth.

Organic mulches that have been on the soil for a while decomposing can provide an adequate growth medium for weeds to germinate and grow in. If crabgrass is germinating in the mulch, move it about with a rake to reduce seedling establishment. Hand pull escaped crabgrass plants before they set seed.

If you're using herbicides, apply pre-emergent herbicides before crabgrass germinates or post-emergent herbicides after it germinates. Avoid using chemical herbicides in vegetable gardens because of the variety of crops grown and planted there. Sure, but grasses are generally not the tastiest weeds out there. That said, crabgrass can be used as a forage crop for livestock and its seeds have historically been harvested as an edible grain.

Lambsquarters (Chenopodium Album)

According to the Weed Science Society of America (WSSA), lambsquarters ranks as one of the most common weeds in North American gardens.

Common lambsquarters is a summer annual broadleaf weed that is widely distributed across the northern half of the United States and southern Canada. Thanks to its widespread distribution, it's no surprise that lambsquarters is often a problem in gardens with sugar beets, vegetable crops, and pulse crops such as dry edible beans, lentils, and chickpeas.

Lambsquarters is a very fast-growing annual with seeds that are small and light enough to be blown by the wind over short distances and can sometimes survive for decades in the soil. Under favorable conditions, these weeds can establish themselves quickly and spread profusely.

How to Control Lambsquarters

This summertime weed rapidly removes moisture from soil, so remove it from unwanted areas as soon as possible. Cultivate lambsquarters out of your garden using a sharp hoe. Yes, you can eat lambsquarters (assuming you're not using chemicals in your garden). In fact, their leaves are quite high in beneficial nutrients. The young shoots and leaves can be eaten raw in any vegetable dish, or sauteed or steamed like spinach.

Pigweed (Amaranthus Spp.)

Pigweed wins the title of most "problematic" annual weed. It has evolved traits that makes it a tough competitor, especially in broadleaf crops like soybeans and cotton.

An annual weed that reproduces by seeds, pigweed is characterized by its fleshly, red taproot. This weed appears in late spring or early summer and likes warm weather.

How to Control Pigweed

Try to pull out this weed before it flowers. Some weed seeds require light for germination and pigweed is one of those. To prevent pigweed in the future, cover your garden plot with a winter mulch. Also, till very shallowly in the spring; only turn up a small amount of soil to keep those seeds buried. When you till you may bring up some pigweed seed so it's best to mulch again. Cover the soil with five layers of wet newspaper and cover that with 3-6 inches of mulch.

Pigweed is also edible—though usually only when young and tender, and when taken from a pesticide-free area. In June, the young leaves of Amaranthus blitum or amaranth are abundant and should be eaten because of their high nutritional content. Vitamin-wise, these greens are packed like carrots and beets and can be delicious in a tossed salad. You can also cook them as you would spinach. Native Americans used the black seeds of this plant as a ground meal for baking.

Chickweed (Stellaria Sp. and Cerastium Spp.)

Common chickweed (Stellaria media) is a winter annual that grows in well-watered areas. It's a reservoir for insect pests and plant viruses.

When growing without competition from other plants, common chickweed can produce approximately 800 seeds and it takes 7 to 8 years to eradicate. Chickseed thrives in moist, cool areas so it often gets started before spring crops can become competitive and can limit vegetable harvest.

Common chickweed is often forms a dense mats and rarely grows higher than 2 inches. The flowers are small with five white petals. Common chickweed will grow in a wide range of soils but does particularly well in neutral pH soils with high nitrogen and poorly in low pH or acid soils.

How to Control Chickweed

Fortunately, annual chickweed is easier to control as long as you pull the weed when the plant is small and before it flowers. The challenge can be locating it during the short period between germination and flower production, so be sure to monitor closely and completely remove the weed so it doesn't reroot. Remember this is a "winter annual." So, monitor the soil surface for chickweed seedlings throughout late fall and winter and then remove them by shallow cultivation or by hand pulling.

Using an organic mulch such as wood chips, at least two inches deep, will reduce the amount of weed seeds germinating by limiting light and serving as a physical barrier. Synthetic mulches such as landscape fabrics may also be used. In landscaped areas, they should be covered with an additional layer of mulch (rock or bark). Vegetable gardens also can utilize black plastic, both as mulch into which seeds or transplants are placed and also between rows. By spring time, we would not recommend chemical controls for this witner annual in the garden. In late fall, consider preemergent herbicides as a last resort. Chickweed is edible. When young, the leaves, stems, and flowers can all be eaten either raw or cooked. It adds a delicate spinach-like taste to any dish.

Dandelion (Taraxacum Officinale)

They provide an important source of food for bees early in the year, too. If you don't mind giving your lawn over to dandelions, that's fine. However, you may wish to consider investing in a lawn. In time, dandelions will also take over any habitat from your garden to your ornamentals to your grasses. They have the most weedy characteristics of all the weeds. Not only do dandelions have wind-borne seed but also reproduce vegetatively thanks to large tap roots. So unless you cut the root deep into the soil, you can rest assured the plant will reemerge.

How to Control Dandelions

Removing dandelions by hand-pulling or hoeing is often futile, unless done repeatedly over a long period of time, because of the deep tap root system of established plants. But if you have a small area, pull young dandelions by grasping them firmly by their base and wiggling gently, as you must dislodge their deep taproot from the soil. Alternatively, use a hand trowel to dig them out. Try to remove the whole dandelion root at once, as any piece left in the ground will probably grow back.

If you keep a lawn, a vigorous (and competitive) lawn will slow down dandelion infestation. Dense turfgrass and ornamentals shade the soil surface, reducing the establishment of new dandelion seedlings. Many broadleaf weeds may be controlled with mowing but this is not true of dandelion. Because it grows from a basal rosette that is lower than a mower blade can reach, mowing will have no effect on control.

For a garden bed, mulches of wood chips or bark are effective if they are maintained at a depth of least 3 inches deep (and replaced over time). Mulching with landscape fabrics can be particularly effective for controlling seedlings, reducing the amount of light that is able to reach the soil. Use a polypropylene or polyester fabric or black polyethylene (plastic tarp) to block all plant growth.

Solitary new dandelion plants along fence rows, roadsides, flower beds, and in turfgrass should be grubbed out (removed by digging out the entire plant, taproot and all) before they produce seed. Dandelion knives and similar specialized tools are available for removing individual weeds and their roots while minimizing soil disturbance. Monitor the area for several months to make sure that removal of the taproot was complete. If you're using herbicides, consider pre-emergence herbicides such as those containing dithiopyr or isoxaben because they are applied to the soil before the seeds germinate.

Yes. The jagged leaves of this perennial are edible, especially when young and tender. The flowers, too, can be eaten raw or fried, or used to make dandelion wine. That being said keep in mind that dandelions are an important source of food for bees in early spring, so you should only harvest a small amount and leave plenty for the pollinators.

Shepherd's Purse (Capsella Bursa-pastoris)

Shepherd's purse is actually a Brassicacae and part of the Mustard family along with cabbage. This flowering annual produces heart-shaped seedpods after flowering. It likes cool weather and its yellowish-brown seeds are long-lived in the ground.

How to Control Shepherd's Purse

Keep an eye out for its distinct leaves and pull out this annual weed by hand before it seeds. Be sure to remove the entire root. The immature heart-shaped seedpods of shepherd's purse have a peppery taste and can be used as garnish in moderation. Other parts of the plant, like the leaves and mature seeds, may cause indigestion and should not be consumed.

Creeping Charlie (Glechoma Hederacea)

Creeping Charlie (ground ivy) and also wild violet are common in shady lawns. Native to Europe, it has become an invasive lawn weed in North America. The plant has bright green leaves with scalloped edges on creeping stems that root at the nodes. It tends to form a dense mat over the ground.

The reason Creeping Charlie is so challenging is the way it spreads—by both seeds and by creeping stems (called stolons) that grow along the ground. If you try to dig it out and leave behind a fragment of rhizome (root), even a tiny piece can grow up as a new plant.

How to Control Creeping Charlie

- Improve turf density by seeding grasses in shady areas which will help to limit this weed from spreading.

- Also, make sure to grow the most suitable type of turfgrass for the location (e.g., plant shade tolerant turfgrass varieties under trees).
- Improve soil drainage or water less frequently to dry the soil.
- Mow regularly (to a height of two to three and one-half inches), fertilizing and watering appropriately, and overseeding in the fall.

Alternatively, consider removing grass and growing shade-loving plants such as vinca, English ivy, pachysandra, or hosta that compete well with weeds (though they can also become weeds themselves, so plant at your own risk.). In areas where Creeping Charlie has become established, try removing plants by hand. This is the control method of choice in vegetable or flower gardens. Try to pull the weed without breaking it and over time it may give up.

However, this may not be a viable option in heavily infested areas, as the extensive spreading stems of creeping Charlie can be difficult to completely remove. If you have mats of weed, smother with newspaper or tarp. Once plants are pulled, make sure to dispose of the plants in such a way that they cannot re-root. Common herbicides do not work.

Prior to the mass cultivation of hops, Creeping Charlie was historically used in the brewing process of beer. As a member of the mint family, it has a slightly minty flavor and is often used by medical herbalists.

Weed Science

Weed science is the study of vegetation management in agricultural production systems, natural areas, and managed properties in residential and urban areas. It involves the development of knowledge in a variety of areas, including weed biology, taxonomy and ecology, biological control, cropping systems, integrated management systems, conservation tillage, transgenic crop tolerance, precision technology, and herbicides. The development of knowledge of the major mechanisms of weed survival is essential for developing effective methods of weed management.

Weed science has made major contributions to increasing crop production over several decades. Herbicides have been the primary control technique. Because of their efficacy and ease of use there has been an overreliance on them at the expense of other weed-control methods. If the only or primary goal is to increase production, then the quest for better herbicides must continue. If the goal is sustainable weed management in a sustainable environment and society, then other control techniques must be investigated. Farmers must be shown that adopting non-herbicidal weed management will not increase the risk of crop failure or reduce profit. The new goal should be development of successful weed-management systems that have minimal or no effect on the flora and fauna of soil, water, or air and have no adverse effects on people.

Weed science, as an integral part of agricultural production needs to evolve by moving away from its mono-disciplinary perspective at targeting weeds, sometimes a single species, through the overreliance on few single herbicide mechanisms of action. Herbicides remain a simple and cost-effective way to control weeds but they are rapidly losing their effectiveness due to evolution of herbicide resistance. Additionally, weed science has been left wanting for a strong theoretical foundation rooted in evolutionary and ecological disciplines therefore, there is a great need for a new weed management paradigm in modern agriculture based on ecological principles and non-conventional weed management approaches. The "many little hammers" concept and the "use of technological advancement" are two major integrated weed management components that are gaining momentum. Automated, robotic weed control is being rapidly developed, particularly for vegetable crops and organic agriculture. Cover crops and weed seed destruction techniques are becoming popular with growers. In the future, RNAi technology, gene editing and robotics will yield new tools for weed control. Agriculture is also moving into a new era of big data or "digital farming." It will be interesting to see what new unforeseen weed control solutions will be derived from this new farming approach that will allow more intelligent application and integration of weed management technologies.

Crop Weed Competition in Field Crops

There are about 30,000 plant species identified as weeds. Among these, 250 are listed as very troublesome in crop production which is an important factor that causing major yield losses. In general, 45 per cent of yield loss caused by weeds (anonymous, 2010) if there is increase in weeds population in different field crops like in rice 69 per cent yield loss, 34 per cent in wheat, 50 per cent in pulses, 72 per cent in sugarcane and in almost all vegetables around 90 percent of yield will reduces. A large gap is noticed between the potential yield and the actual yield obtained and globally 287 tonnes of food loss due to weed infestation which accounts for 11.5 per cent of the total food production.

So that weed management is one of the important strategies for minimising the yield loss. Before taking management strategies we should know their biology like competition themselves and with crops. Weeds and crops are very similar in their life cycle for rooting, growth and development, efficient utilisation of resources available their surrounds. Both weeds and crops are competing for carbon-di-oxide and nitrogen from the atmosphere, water and minerals from the soil and light from the sun for their growth and development. As a result, growth of the crop plant is restricted and yields are drastically reduced. The yield losses due to weeds have been reported to vary from 16 to 68 per cent in cultivated crops under different agro-climatic conditions.

A concept of competition is the struggle for survival and continued for existence. In similar way crop weed competition is the relationship between two or more species in which supply of growth factor falls below their combined demand. The competition does not occur when the growth factor is abundant. However, it starts immediately when growth factors fall short in supply. In similar way, the critical period of weed competition might have originated from the belief that weeds are not equally damaging throughout the crop period. There may be a certain stages in crop growth period when weeds are more harmful to crop growth and yield. Because of its initial slow growth weeds are taking advantage to utilise more resources and dominate over crops. Mainly weeds compete with crop for nutrients, solar radiation, soil moisture etc. Weeds compete with crop whole life cycle but its effect does not remain same during all stage of crop growth. Therefore the short time span in the life cycle of crop growth, when weed causes maximum reduction in its yield or in other words, when weed control measure if adopted may fetch near maximal or maximum acceptable crop yield it is known as critical period of crop weed competition. It is therefore, simply the specific duration of weed free situation of a crop resulting into near maximal yield, which is sufficiently close to that obtained by the season long weed free situation.

Critical Growth Period of Weed Control

The critical period of weed control (CPWC) is the period of crop growth when the crop must be kept weed-free to prevent yield loss due to weed interference. The Critical period for weed competition is estimated from two factors, namely, (1) the critical weed-free period (CWFP) defined as the minimum duration of a weed-free period required following planting to prevent yield losses above 5% threshold and (2) the critical time of weed removal (CTWR), the maximum period of time a crop can be exposed to early season weed competition before it reached to threshold yield loss The CTWR and the CWFP delineate the beginning and end of the CPWC, respectively. Similarly, results show in lentil crop the CPWC studies have been quite variable. Researcher showed that between 30 and 60 days after emergence (DAE), between 60 and 90 DAE and between 49 and 56 DAE was the CPWC in lentil in India, Syria and Jordan, respectively. In similar way, researcher estimated between 38 and 92 DAE in Jordan based on a 10% yield loss threshold and 5% yield loss threshold, is between 2 and 4 or 4 and 6 weeks after emergence depending on location in Iran.

In western Iran Mashhadi and Ahmadi estimated a CPWC of 27 to 44 DAE, between the 6 and 14 leaf stage, and 205 to 385 GDD. Researcher in Jordan estimated 35 to 49 DAE; Researcher in India estimated 0 to 56 DAE and 28 to 56 DAE, respectively; and Researcher in India estimated that hand weeding at 30 and 60 DAE would prevent unacceptable yield loss from weeds.

Weeds Effect on Physiological Parameters of Crop

A competition between among and between the species which may strongly changes their physiological characteristics of growth and development. It leads to differences in the use of environmental resources, especially the water, which directly affects the availability of CO_2 in leaf mesophyll and leaf temperature therefore, the photosynthetic efficiency decreases. According to researcher the chlorophyll (chl) content (SPAD value) was decreased with increasing the duration of weed interference period. The maximum chl content (42.10) was observed in the season long weed-free treatment followed by 75 day weed-free and 30 day weedy treatments (>41) while the minimum chl content was found in the season-long weedy treatments. Chlorophyll is the main pigment of photosynthesis in plants. It is strongly influenced by environmental factors. Researcher observed the significant reduction in leaf chlorophyll content in sugar beet from the competition with Chenopodium album and chlorophyll content became reduced with increasing weed competition and weeding durations are significantly influenced with chlorophyll content of cowpea.

Researcher shown that the relative growth rate (RGR) was highest between 6 and 8 WAS in MK 373 and 8-10 WAS in Samnut 10 of two groundnut variety and declined with advent of time thus all the weed control treatments had higher RGR as compared to the weedy check.

Crop Weed Competition

Crop weed competition indicates the competition between crops and weeds in natural eco-system in response to resources struggle for their existence and superiority. Mainly the crops and weeds are competing for sunlight, CO_2, soil moisture, nutrients, space etc. to complete their life cycle. These points explained in following heads.

Competition for Light

Under thick canopy light cannot be stored like water or nutrients and hence must be intercepted as and when it is available or else it will be lost forever. Weeds deplete Photosynthetically Active Radiation by shading of lower leaves. Researcher reported that the seed yield increased in narrow row to wide row cultured soybean and this attributed to greater light interception. Researcher observed a linear relationship between weeds and soybean and showed that the variation accounts for 86 per cent due to shading by the weeds. It was predicated that 19 to 25 per cent yield loss was observed due to 44-56 per cent shading of the crop by the weeds. Researcher reported that, soybean yield losses were highly correlated (r^2= 0.84) with leaf area of weeds as viewed from directly above the weed crop canopy. Further, they stated that after a week of soybean emergence, the weed canopy diameter measured from overhead photographs also correlated well with soybean losses (r^2 = 0.82), but correlation with actual leaf area of weed was not significant (r^2= 0.31) Berti and Sattin was of the opinion that the competitive effect of the weeds appeared to strictly related to other relative cover, indicating that for weeds growing taller than crop, the main competitive factor may be the shading caused by the leaves of weeds situated above the canopy.

Competition for Nutrient

Researcher reported that Uptake of N, P and K by parthenium increased with increase in its competition period. This might be due to higher accumulation of dry biomass per unit area. Nitrogen, phosphorus and potassium uptake by parthenium was in the range of 2.7-18.4, 0.2-2.4 and 2.3-17.7 N, P and K kg ha^{-1}, respectively, at its different competition periods. Parthenium N uptake response to increasing density was quadratic during both experimental years. However, both the P and K uptakes showed a linear trend and quadratic trend during 2012 and 2013, respectively. Increase in NPK-uptake by parthenium with increasing crop-weed competition duration might be the result of greater plant biomass accumulation with prolongation in its growth period. Researcher were also reported that N uptake by velvetleaf (Abutilon theophrasti), a broadleaf weed in maize increased significantly in linear as well as quadratic pattern by increase in competition period from 0 to 100 days after crop emergence.

Researcher indicated that N, P and K contents in black seed grains and tissues significantly decreased as the weed-black seed competition increased during WI periods. Overall, N, P and K contents in weed species tissue were found to be 1.8 to 2 times higher those that of black seed. A significant decrease in N, P and K contents in black seed tissue emphasize the poor absorption ability of nutrients by black seed in competition with weed species. Researcher observed high uptake of N, P and K by weeds under weed-cumin (Cuminum cyminum) competition. Researcher reported that weeds in groundnut crop accumulated 49.50, 8.49 and 59.47 kg N, P and K ha^{-1}, respectively, at harvest.

Competition for Moisture

The competition for water occurs in below ground it is mainly depends on rooting pattern and their volume. For producing of equal amount of drymatter the weeds absorbs more moisture from soil than the crop. In pearl millet field the cynodon dactylon has twice the water transpiration than pearl millet because of its roots are more deeper in soil. In weedy field the moisture is more extracted by weeds and crops are at plants reaches active vegetation to reproduction stage. From water use efficiency point of view, most of weed species are C_4 type having higher WUE as compare to the C_3 type of crop species.

Effect of Crop Weed Competition on Growth and Yield Attributes

Plant Height

Researcher observed that rice plant height and weed competition period was significantly influenced by increased the length of weed interference were but there was no significant adverse effect after 45 DAT in weedy treatments. Azmi pointed out that the plant height of rice decreased when weeds were allowed to compete till 30 DAT and up to harvest. Begum observed that the plant height of rice significantly reduced when rice plant competed with F. miliacea up to 70 days or longer. Similarly, Chauhan and Johnson noted that plant height of rice was significantly reduced by competition with wild rice (Echinochloa colona), and the reduction was increased in higher weed density. Researcher reported that rice plant height was significantly decreased with weed competition up to 40 days or longer. Researcher indicated that black seed height was strongly affected by different Weed Free (WF) and Weed Interference (WI) periods. Entire growing season of black seed in weed interference treatment caused to an increase in height by 30.3%, as compared with the WF throughout the growing season treatment. And higher weed height, especially C. album, Echinocloa crussgalli and A. retroflexus compared with black seed plants (approximately 1.2 to 2.6 times); therefore, probably there was more competition between black seed and weed community over light absorption.

Dry Matter Accumulation

The highest weed DM was found with season long weedy plots. The treatments where weeds were allowed to grow at the early stages (15 DAT) accumulated higher DM than those allowed to grow at the later stages of the crop growth in rice. The weeds emerging at the later growth stages offer less competition to crops as it accumulates lower DM.

Leaf Area Index

Researcher reported that with increase in the length of weed interference period rice leaf area was adversely affected and, as compare to increasing span of weed free period, up to 30 DAT. The maximum leaf area was observed in the season-long weed-free treatments with 1185.6 cm2 hill^{-1}, and gradually decreased in all the weedy treatments. Similarly, DM production in rice was decreased with the duration of weed competition. Season long weedy treatment produced lowest rice plant biomass due to a consequence of disturbance in mineral supply, lower water potential and nutrient uptake disturbance by weeds which resulted in reduced growth and lower rice leaf area. Effective weed control is necessary to enhance LAI of groundnut. Researcher observed that in Sourashtra region of India, LAI of bunch varieties of groundnut might be 1.7 at 60 days after planting (DAP), and might increase to 4.0 at 90 DAP. LAI of groundnut was reached 3.0 at 64 DAP, which at maturity (137 DAP) were reduces to 1.7. The reduction in LAI at 12 WAS in Samnut 10 and 10–12 WAS in MK 373 was due to pest attack). Reduces the corn leaf area with Weed interference by reducing the expanded leaf area of each individual leaf and accelerating senescence of lower leaves.

Yield Performance

Researcher reported that the maximum value of yield attributes (viz., total number of developed pods, hundred pod and hundred kernel, test weight, shelling percentage and volume weight) were observed in weed free treatment followed by two hand weeding and hoeing at 15 and 30 DAS and pre-emergence pendimethalin followed by one hand weeding at 30 DAS. Similarly, Researcher observed the highest number of matured pods per plant, seed weight per plant, 100 seed weight, pod yield, seed yield and harvest index were recorded in rice straw mulch + one hand weeding at 6 WAS over all other treatments. Lowest yield components and yield were recorded in weedy check. Weed competition throughout the crop duration resulted in 100% yield loss in both rice cultivars compared to weed-free conditions, in which yield was 6.39-6.80 t ha^{-1} for cultivar PR 114 and 6.49-6.87 t ha^{-1} for PR 115. The weed free control had the highest ear weight (260 g), 1000 kernels weight (356.67 g), the number of kernels in ear (785.33), biological yield (56800 kg ha^{-1}) and grain yield (16660 kg ha^{1}).

The full season long weed competition treatment had the lowest ear weight (160 g), 1000 kernels weight (236.67 g), the number of kernels in ear (380.5) biological yield (27400 kg ha^{-1}), grain yield (5525 kg ha^{-1}) and the harvest index (20.36 %). The highest harvest index was achieved in weed control up to 10 days after emergence in maize. In case of fababean grain yield and yield contributing traits were significantly affected by weed competition. Weed-crop competition may end at 45 days. Grain yield losses due to uncontrolled weed growth throughout the crop cycle were 46%. At the same time, plant height, numbers of pods per plant, numbers of seeds per pod and 1000 seed weight were significantly decreased due to weeds. Researcher showed, yield attributes and grain yield declined with the increased duration of crop-weed interference period and increased with long weed free durations in spring maize.

The highest yield of groundnut pods (16.18 q ha^{-1}) and haulm (10.30 q ha^{-1}) were observed in weed free check. Reduction in crop yield had direct correlation with weed competition. In drought situation, weeds thrive better than crop plants. When left uncontrolled, weeds can grow taller than crop plants and suppress the crop growth. Yield reduction from 30 to 35 per cent due to the presence of weeds had been reported in groundnut. The yield loss caused by weeds was estimated at 30 to 50 per cent in groundnut. Researcher reported yield reductions as high as 90 to 95 per cent due to weeds depending on the type and density of weed flora.

Study on Weeds

Increasing weeds population in specified area use to utilise more resources efficiently and build their growth and development and less availability to crops in same space. Weeds are grasses, broad leaved weeds and sedges are different competing ability with crops. For better weed management need to know the different weed floras, weed density and its dry matter holdings.

Weed Flora

In general, around 40% of crop yield loss by weed infestation. It mainly depends on the type, species and density of weeds growing in a crop community, as weeds vary from place to place and season to season. Researcher reported that dominant weed species in rice field during both years were Cyperus rotundus and Cyperus compressus among sedges; Echinochloa crus-galli, Echinochloa colona, Dactyloctenium aegyptium, Digitaria ciliaris, Eragrostris spp. among grass weeds; Euphorbia hirta, Phyllanthus niruri, Trianthema portulacastrum, Ammannia baccifera among broadleaved weeds. Similarly, Rottboellia cochinchinensis, Digitaria ciliaris, Echinocloa colona, and Eleusine indica. Ageratum conyzoides, Amaranthus spinosus, A. viridis, Commelina benghalensis, Corchorus olitorius, Cyperus iria, Cyperus rotundus, Dactyloctenium aegyptium, Eclipta prostrata, Portulaca oleracea, and Trianthema portulacastrum were others species common during both seasons. Researcher were reported that the major weeds associated with sunflower on sandy loam soils of Kanpur were, Cyperus rotundus, Spergula arvensis, Anagallis arvensis, Melilotus sp., Convolvulus arvensis, Asphodelus tenuifolius, Launaea asplenifolia, and Chenopodium album. Researcher reported that Dominant weeds associated with the sunflower crop were Euphorbia sp. Amaranthus polygamous, Sonchus arvensis, Digera arvensis and Denebra sp. Similarly, researcher reported that Dinebra retroflexa, Phyllanthus maderaspatensis, Digera arvensis, Acalypha indica and Cyperus rotundus are the dominated weeds in sunflower field.

Weed Density and its Dry Matter Production

Researcher observed that in black seed (Nigella Sativa L.) both years of the experiment, the highest weed densities were observed at 42 days after emergence and then declined. On 42 days after seedling emergence, total weed density was found to be higher in the first year (384 plants m^2) than in the second year (312 plants m^{-2}). However, during both years, dominant weed species were very similar. Researcher parthenium weed dry mass increased in a quadratic fashion with increase in its competition period. The higher dry biomasses (32.9 and 33 g m^{-2}) were gained by full crop season parthenium weed competition during years 2012 and 2013, respectively. The less parthenium weed dry biomasses (6 and 8.2 g m^{-2}) was recorded in plots with parthenium weed compete with maize up to 35 days after crop emergence (DAE) during 2012 and 2013. Increase in dry biomass of parthenium with extended competition duration was presumably attributed to prolonged growth span of weed ultimately resulting in more biomass accumulation.

Harmful Effects

Reduction in crop yields due to weeds result from their multiferous ways of interfering with crop growth and crop culture. Weeds compete with crops for one or more plant growth factors such as mineral nutrients, water, solar energy and space and they hinder crop cultivation operations.

- Competition for mineral nutrients: Being hardy and vigorous in growth habit; they soon outgrow the crops and consume large amount of water and nutrients. Thus causing heavy losses in yields. For e.g. Mani, found that on an average weeds growing in crop field during the kharif season removed 46.6 kg/ha N, 12.1 kg/ha P and 73.3 kg/ha K. In general, weeds removed N and K from soil in much greater quantity than P.
- Competion for water: For producing equal amount of dry matter, the weeds, in general transpire more water than most crop plants. It is reported that wild mustard transpires about four times more water than a crop of oat.
- Competition for Solar Energy: About 99% of dry matter in plants is made up of organic matter that is dependent on solar energy. When plants are mutually shaded, their production

potential is greatly reduced even though water and other nutrients are available to them in abundance.

- Competition for space: Weeds compete for space both in the rhizosphere and atmosphere. In the presence of weeds, crop plants also have limited space to develop their shoots, which amounts to reduced photosynthesis in them.
- Weeds reduce the crop quality: Weeds may reduce the quality of the crop produce in many ways. Weed seed like wild mustard, sweet clover, a Mexican poppy and bulblets of wild garlic and wild onion when threshed and ground with winter grains can results in serious consequences besides imparting objectionable odour to the flour. Khanna observed that striga (striga Spp.) reduced the quality of sugarcane juice by 3.9 to 8.9 percent.
- Weeds impair the quality of the animal produce: Many weeds in pastures and forage crops impart undesirable flavours to milk and meat of animals. For e.g. Pivali tilwan (Cleome viscosa) imparts undesirable flavour to milk. Gokharu or Landaga (Xanthium strumarium) get attached to the body of sheep and seriously impair the quality of wool.
- Weeds harm animal health: Several weeds of grasslands and forage crops contain high alkaloids, tannis, oxalates, gulcosides, and other substances that prove poisonous to animals when ingested. For e.g. Silky lupine (Lupinus sericeus) is responsible for crooked calf disease.
- Weeds harbour insect pests or diseases: Weeds either give shelter to various insect pests and diseases or serve as alternate host. For e.g. Weed around paddy bunds harbour the gallfly.
- Weeds damage human health: Health, comfort and work efficiency of man are also affected by weeds directly or indirectly. For e.g. people in U.P. are plagued year after year with hay fever and asthma aggravated by pollens of regweeds bursage. Tsetse fly which cause African sleeping sickness.
- Weeds contaminate water bodies: Aquatic weeds change the flavour appearance and taste of drinking water. Aquatic weeds are a menace to fisheries too. Aquatic weeds on decomposition gives offensive odours and pollute atmosphere.
- Weeds cause quicker wear and tear to farm implements: Being hardy and deep-rooted, the tillage implements get worn early.
- Weeds reduce the value of the land: Agricultural lands heavily infested with perennial weeds like Kans (Saccharum spontaneum) always fetch less price.
- Less efficient use of land: In case of perennial weeds, the carrying capacity of the grazing lands is reduced and cause depreciation of land value.
- Increase in cost of cultivation: In fields of crops infested with weeds, the tillage operations require high cost.
- Disturbance in Public places: It is desirable that public places be kept clean of weeds. Presence of weeds around our living and working places makes the surroundings dull.

References

- Introduction-to-weeds-what-are-weeds-and-why-do-we-care: psu.edu, Retrieved 02, July 2020
- Pictures-of-weeds-4069715: thespruce.com, Retrieved 25, January 2020
- Weed-science, agricultural-and-biological-sciences: sciencedirect.com, Retrieved 08, May 2020
- A-study-on-crop-weed-competition-in-field-crops-343651429: researchgate.net, Retrieved 22, August 2020
- Harmful-effects, weed-control: indiaagronet.com, Retrieved 05, March 2020

Chapter 2

Classification of Weeds

There are various criteria on the basis of which weeds can be classified such as anatomy, cotyledenous character, nature of stem and association. On the basis of anatomy, weeds can be grouped into annual, biennial and perennial weeds. This chapter has been carefully written to provide an easy understanding of the classification of weeds.

Classification According to Anatomy

Annual Weeds

Annual weeds are the ones that last for one year or growing season only, but which produce lots of weed seeds for next year - and years afterwards.

Fat Hen (Chenopodium Album)

The Fat Hen weed is the fastest-growing of all of the annual weeds. There are many different varieties. It is possible for a Chenopodium Album to produce 20,000 seeds in its lifetime just from one single plant. In its baby form, the Fat Hen can be used in salads as it is very tasty but must be monitored carefully in order to prevent the growth across open ground.

Young fat hen weeds have bright green leaves which tend to darken with age and look similar to the image above. The chenopodium album will grow between 2 and 4 feet tall with a vertical stem. This can be difficult to identify for beginners as it can appear in many intermediate forms but the colour and shape of the leaves is the main identifying trait. The lower leaves are usually shaped like a rhomboid whereas the upper leaves would usually appear as a branched stem.

Chickweed (Stellaria Media)

Chickweed is one of the most common in the UK. It varies greatly in size and weight and can be found in a multitude of locations including your garden. Like the Fat Hen, the Chickweed's baby form is very tasty and makes a great addition to a salad, as well as being loaded with vitamins. You are advised to disperse of this weed as soon as possible as it attracts Spider Mite and Greenfly which will ruin any edibles you are growing.

Chickweed is most easily identified by its small white star-shaped flowers and tiny white line of weaved hair that grows along the stem. You can expect the stellaria media to vary in size between 5cm and 50cm with oval-shaped leaves that have pointed tips. The texture of the leaves can vary from smooth to slightly hairy.

Groundsel (Senecio Vulgaris)

This weed is capable of being blown in by wind from surrounding areas such as fields or neighbouring gardens, so don't be surprised if it crops up at random. Like most weeds, Groundsel can produce a large amount of seeds, even to the extent that it can produce numerous generations of weed within one growing season.

A groundsel weed is capable of growing up to 2 feet tall with a variety a leaf ranging from lightly hairy with cotton-like hairs to smooth. The stems branch out into little clusters which have small yellow flower heads. The leaves of the Senecio vulgaris are evenly spaced on the stem and have toothed edges.

Hairy Bittercress (Cardamine Hirsuta)

A fairly common little weed in the UK often found in areas of bare soil and pathways. Whilst it only grows to a few centimetres high, this little rascal has a hidden talent. The seed pods formed after flowering can explode and disperse the seeds several feet away, so make sure you identify and control this weed well before its flowers set to seed.

Annual Meadow Grass (Poa Annua)

One of the most common coarse grasses found growing in lawns, flower beds, paths etc. This grass weed is capable of flowering even when the lawn is mown short and the small seed heads can quite easily be seen creating a speckled effect. Annual meadow grass does not like acidic soils, so avoid liming your lawn if you have this grass weed. Sulphate of Ammonia will increase the soil acidity but only use sparingly as other lawn grasses will not like a heavy dose of this.

Yellow Oxalis/Creeping Woodsorrel (Oxalis Corniculata)

A low-growing annual weed with bright yellow flowers. The clover-like appearance of the leaves can sometimes lead to mistaken identity. With the stems being able to root where they touch, it doesn't take this ground invader long to cover a fair-sized area, along with the addition of producing a vast quantity of seed which is thrown out of their seed pods to ensure an even wider spread.

Control

Annual weeds have very simple and shallow root systems, so simply hoeing the tops off or pulling the weeds up will be enough to control them. The roots simply don't have the ability to gather enough energy to re-grow from scratch.

Contact weedkillers, usually marketing as working within a few hours, will also be an option. They burn off the top growth of the weed quickly and are often referred to as 'chemical hoes'. Contact weedkillers don't get absorbed into the roots, so they are not the best choice for tough perennial weeds.

Whichever method you choose, make sure you remove the weeds before they set seed. Annual weeds spread by sending out vast quantities of seeds so you don't want to let the weeds flower for long.

Biennial Weeds

This type of weed grows in two seasons. In the first season they remain in a vegetative state and rarely flower or set seeds. In the second season, the biennial weed will flower, set seeds, and then typically dies. Proper maintenance of grass and lawn care can prevent the occurrence of biennial weeds which can take nutrients and water from grass or other plant life. Biennial weeds are most commonly found in no-till fields, pastures, and fencerows that are not mowed. They are easiest to control in the seedling stage. A rosette is a plant form with no central stalk. All leaves in a rosette arise from close to the soil surface, as in thistle. Many familiar weeds and wildflowers are biennials. Queen Anne's lace (wild carrot), evening primrose, burdock, common mullein, and moth mullein are examples of biennials found in area. Biennials can have large taproots, which sometimes causes people to confuse them with perennials. Burdock and Queen Anne's lace (wild carrot) are biennials commonly found in New England landscapes. There are few turf weeds which are biennial. Biennials have taproots that must be killed or removed for lasting control. This can be done by hand or with a translocated postemergence herbicide. Growth of these weeds can be prevented with mulches. Biennials spread only by seed and are less likely to be serious problems than many perennials.

Perennial Weeds

Perennial weeds are plants that live more than two years. Perennials reproduce from seeds and vegetative parts, such as creeping roots, rhizomes, bulbs, tubers (below-ground root-like stems), or stolons (above-ground stems that produce roots). Common perennial weeds in rice are Paspalum distichum, Cynodon dactylon, Scirpus maritimus, and Cyperus rotundus.

Management Options

Land Preparation During the Dry Season

Perennial weeds reproduce effectively through vegetative means. Cultivating the soil at the harvest of the last crop and occasional disturbance of the soil during dry season will expose the vegetative propagules to sunlight and are killed by desiccation. This method works well against C. rotundus and C. dactlyon in rainfed areas where fields are left fallow during the dry season.

Land Preparation and Flooding

In irrigated areas, long land preparation duration of about a month using proper tillage intervals can reduce the population of P. distichum. Flooding the field 2-3 days after transplanting and before the propagules sprouts can reduce the population P. distichum and C. rotundus.

Manual and mechanical: Manual weeding or mechanical weeding (use of rotary weeder) plus handweeding can control perennial weeds however it is costly.

Chemical Control

Bentazone or 2,4-D or the combination of both effectively controls S. maritimus when applied at early stages of growth.

Glyphosate also controls perennial weeds. It should be applied to actively growing seedlings of perennial weeds. As pre-planting treatment, after cultivation, allow the weeds to grow for at least 2 weeks, apply the herbicide and allow at least 1 week after application before the next cultivation to allow glyphosate to fully take effect.

Crop Rotation

Rotating flooded rice with dryland crop can help reduce the population of perennial weed such as S. maritimus. Changing the soil condition from flooded to dryland (aerobic) limits the reproduction of S. maritimus because of the absence of water.

Classification According to Cotyledonous Character

Monocot and Dicot

A cotyledon is an embryonic leaf in seed-bearing plants. There are usually one or two cotyledons for most weeds and the cotyledons either contain or have access to stored food the seedling will need before it can produce its first true leaves. Plants with one cotyledon are called monocots and plants with two cotyledons are called dicots. In most cases the cotyledon in monocots, for example corn, will stay at or below ground level. When corn is planted, the kernel of corn doesn't come up out of the ground, it stays below the soil surface. The cotyledon transfers the energy stores in the endosperm of the seed to the growing plant. Conversely, the cotyledons in dicots, a green bean for example, not only serve as energy stores for the new plant but will push up out of the ground and photosynthesize before the first true leaves emerge.

The difference in the number of cotyledons is only the beginning of the differences between these two types of plants. Differences include venation patterns, vascular bundle arrangements, roots systems and flower anatomy.

In monocots like grass, corn, or daffodils, the veins in the leaves run parallel to each other along the length of the leaf or stem. The vascular bundles are usually arranged in a complex pattern and the root system is pretty fibrous. The flower structures in monocots are arranged in multiples of three.

In dicots like green beans, woodsorrel, and most woody trees and plants, the veins in the leaves form a complex, netlike system. Instead of running from one end of the leaf to the other, the veins will branch off one central vein to form a network. The vascular bundles in dicots will usually be arranged in a ring. These rings are most easily seen in trunks of trees. The root systems in dicots will usually have a primary taproot. Of course plants adapt as well as they can to the site they're planted in, so dicots planted in extremely compacted soil may not have a well-developed taproot. Lastly, the flower structures in dicots are arranged in multiples of four or five.

Growing Points

The last obvious difference between monocots and dicots is the location of the growing points. In monocots, the growing point is at or just below the soil surface and is often protected by a sheath. In dicots, there are multiple growing points which are located at the end of every stem.

The different location in growing points is what allows turf weeds to be managed by consistent mowing. The growing point in turf is at the soil surface and is not damaged by consistent mowing. However, the growing points on weeds like bittercress or wood sorrel are above the soil surface. Consistent mowing means consistent removal of growing points for the weed.

Pesticide Use

Understanding the difference between monocots and dicots is also important when selecting a pesticide as part of an integrated pest management program. Non-selective herbicides aren't usually the best choice when managing weeds in a turf stand, unless of course the intent is to wipe out all things green and begin again. Turf managers who need to manage broadleaved weeds (dicots) within their turf stand will want to apply a selective herbicide. 2,4-D is a great example of a selective herbicide. While applied to all vegetation in a turf stand, grasses inactivate 2,4-D while broadleaf dicots do not, thus killing the dicots while leaving the turf unharmed (when applied within label rates).

However, landscape managers will want to avoid spraying 2,4-D on planting beds as the herbicide will harm the desired plant. Landscape managers will want to select an herbicide that will kill the monocot weed (usually grass) such as sethoxydim to control grasses that are invading the planting area.

Broadleaf and Grass Weeds

Weed scientists commonly speak of broadleaf and grass weeds, which correspond approximately to the Dicot and Monocot subclasses, respectively. This is a useful characterization, not only in conventional weed control (grasses and broadleaf species have different sensitivities to different chemical herbicides), but also in organic weed management. Grasses and broadleaf weeds respond differently to cultivation, mulching, cover crops, and other practices.

Strictly speaking, the Monocot weeds include sedges, rushes, dayflowers, and wild relatives of onion, garlic, and lilies, as well as true grasses, and are collectively called grass-like weeds. In addition, a few kinds of non-flowering plants can become weeds. These include field horsetail (Equisetum), mosses, and liverworts, which occur in moist environments in orchard and landscape, turf, and greenhouse container plantings, respectively. They are rarely major weeds in cropland, and will not be discussed further here.

Amaranthaceae—The Amaranth or Pigweed Family (Dicot)

The amaranth family includes several major weeds of vegetables and other cultivated crops. These include smooth pigweed and spiny amaranth, cited as the world's 14th and 15th worst, as well as redroot pigweed, Palmer amaranth, Powell amaranth, common waterhemp, tall waterhemp, tumble pigweed, and prostrate pigweed, which are common in the United States. Well-adapted to high temperatures, the pigweeds are among the most aggressive of summer annuals and can cause substantial yield losses, even in vigorous crops like sweet corn or tomato. Palmer amaranth and the waterhemps have become particularly troublesome in the southern and midwestern United States because of their rapid development of resistance to glyphosate and their large size. Some amaranths are cultivated for their highly nutritious greens, edible grain, or ornamental flower heads.

A large pigweed plant with seed heads approaching maturity.

Amaranths typically have dense spikes of minute flowers, tiny seeds, simple leaves, and sometimes a distinct reddish coloration of the taproot and lower stem. A single mature pigweed can shed 100,000 to 1 million viable seeds.

Apiaceae or Umbelliferae—The Carrot or Umbel Family (Dicot)

Figure shows the flowers in the Apiaceae family are arranged in simple or compound umbels (a). Individual flowers have a nectiferous disk, which attracts natural enemies of many insect pests. Wild carrot or Queen Anne's lace (b-d) is a well-known member of this family. During its first year of growth, this biennial forms a basal rosette (b), in contrast to the taller flowering form it takes on in its second year (c). The maturing seed head is shown in (d).

This plant family includes two highly poisonous weeds—poison hemlock and water hemlock—which can be life-threatening to humans or livestock if ingested. The other widely-known weed in this plant family is Queen Anne's lace, or wild carrot, a Eurasian species from which the modern vegetable carrot was derived through breeding and selection over the past 2,000 years. Other horticultural crops in this family include celery, celeriac, parsnip, parsley, dill, fennel, cilantro (coriander), caraway, bishop's weed, and many ornamentals. Most members of this plant family provide excellent nectar sources for natural enemies of some insect pests of vegetable crops, and thus have significant economic value in organic insect pest management.

The key characteristic of the carrot family is the umbel, an inflorescence in which all of the individual flower stalks arise from a single point. Some species have compound umbels. Other characteristics include highly divided or dissected leaves that may be aromatic, and small, shallow, nectar-rich flowers. Most species in this plant family have a biennial life cycle. Wild carrot sometimes occurs in vegetable fields but more often in pasture or hay fields, and frequently in field margins, where some farmers welcome it as beneficial insect habitat.

Asclepiadaceae—The Milkweed Family (Dicot)

Milkweeds most often occur in field margins and pastures. However, common milkweed can become a weed of vegetable and field crops, especially when reduced tillage or no-till practices are implemented.

Plants in the milkweed family are characterized by the milky sap that exudes from severed stems or leaves. Common milkweed is an invasive perennial that propagates through a deep, extensive network of rhizomes, as well as seed. The plant forms tall (up to 6 ft), sparsely-branched stems with opposite, simple leaves, and globular inflorescences or flower heads consisting of up to 100 or more individual flowers.

Common milkweed seedling (a), vegetative shoot emerging from rhizome or rootstock (b), mature plant in flower (c), and close-up of inflorescence (d).

Asteraceae or Compositae—The Daisy or Sunflower Family (Dicot)

Many members of the large and diverse plant family Asteraceae have become serious agricultural weeds, including common ragweed, Canada thistle, musk thistle, annual sowthistle, prickly lettuce, spotted knapweed, common cocklebur, common dandelion, galinsoga, common yarrow, and others, many of which frequent vegetable fields. Crops in this plant family include sunflower, safflower, lettuce, endive, artichoke, zinnia, dahlia, and many other cut flowers and ornamentals.

The most definitive characteristic of the Asteraceae is the inflorescence. What we commonly call a daisy or zinnia flower is actually a head, an inflorescence made up of dozens or hundreds of tiny individual flowers packed together hence the old family name Compositae. The readily-accessible pollen and nectar of the composite flowers make both cultivated and weedy members of this plant family excellent food sources for the adult phases of many beneficial insects, such as parasitic wasps and hoverflies.

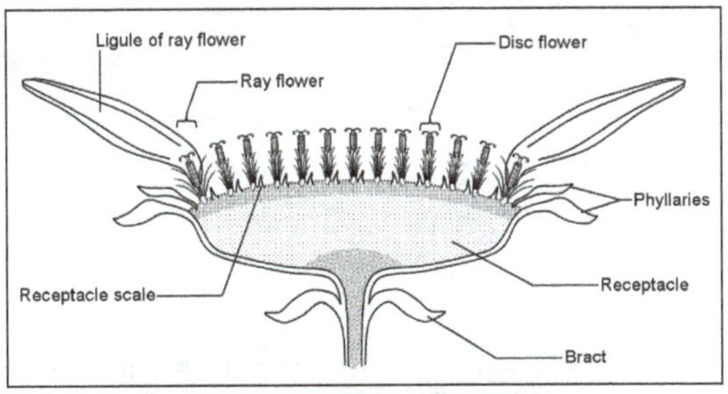

What we call the flower of a plant in the composite family is actually a head made up of dozens or hundreds of tiny individual flowers in "disk" and "ray" forms. Chicory is a common weed in this plant family with showy blue ligulate (ray only) flower heads.

This large plant family includes annuals, biennials, and perennials with or without taproots, rhizomes, or tubers. Many weeds in this family have strongly lobed or divided leaves, which grow in basal rosettes during the vegetative phase of the plant's lifecycle.

Brassicaceae or Cruciferae—The Mustard or Cabbage Family (Dicot)

The mustard family includes a number of common weeds of cool-season vegetables and cereal grains, such as wild mustards, shepherd's purse, and yellow rocket. Garlic mustard is an invasive exotic species of particular concern because its strongly allelopathic properties help it to invade woodlands in the United States. Many of our most important vegetable crops also come from the brassica family, including cabbage, broccoli, cauliflower, kale, collards, bok choy, various cultivated mustards, arugula, and other gourmet greens.

Members of the mustard family have a characteristic flower structure with four petals arranged in a cross, and six stamens. The stamens are arranged in an inner circle of four, and two outer, shorter stamens. Other characteristics include a distinctive seedpod partitioned lengthwise into two valves that split apart from base to apex at full maturity, and accumulation of pungent, sulfurous compounds called glucosinolates, which have allelopathic and antimicrobial properties.

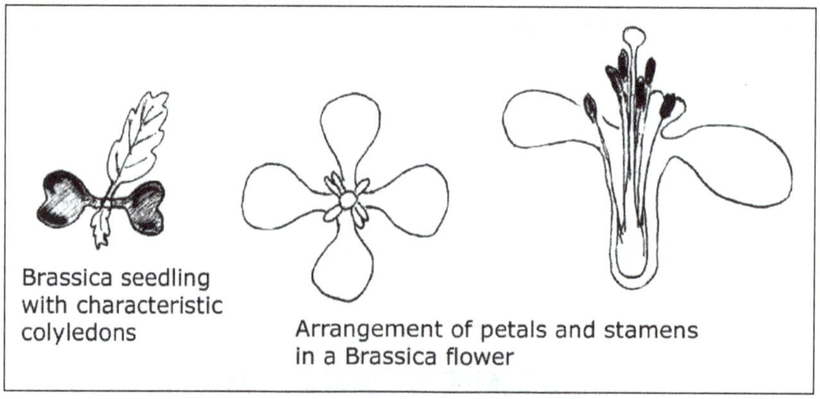

Brassica plants have a characteristic cotyledon shape with a distinct notch at the apex. The flower has four petals arranged in a cross with six stamens. This plant family includes several winter annual weeds such as shepherd's purse, as well as vegetable crops such as cabbage and broccoli.

The mustard family consists of herbaceous plants of cool to temperate climates. While the weedy members of this family are well known and sometimes troublesome to vegetable growers, only one—shepherd's purse—is listed among the world's 76 major weeds. Brassica weeds can be a reservoir of pests and diseases of related vegetable crops, yet flowering brassicas (weedy or cultivated) often attract the natural enemies of the brassica pest complex.

Caryophyllaceae—The Pink Family (Dicot)

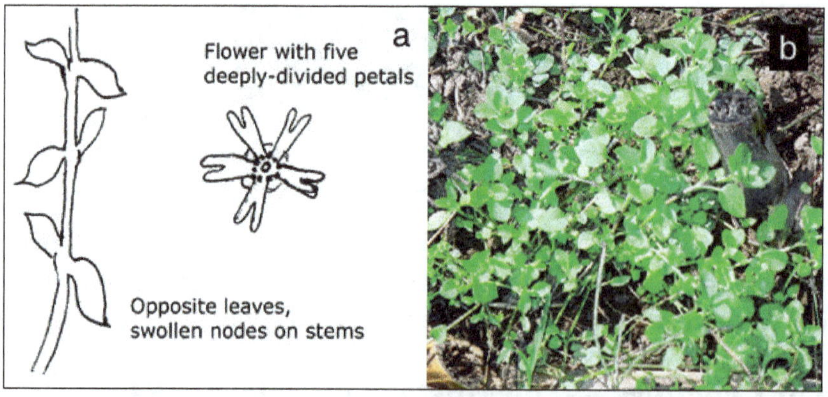

Weeds in the pink family have oppositely-arranged leaves, and small five-petaled flowers with the petals deeply lobed, giving the appearance of 4(a). Common chickweed (b) is a widespread winter annual or occasionally perennial weed in this family.

The pink family includes common chickweed, mouse-ear chickweed, corn spurry, and corn cockle, as well as ornamental flowers such as carnation, pinks, sweet William, baby's breath, and Maltese cross. Common chickweed is particularly troublesome in garlic, early greens, other cool season vegetables, and cool-season forage crops, but it usually does not seriously impact warm-season vegetable or row crops.

The pink family is characterized by simple, opposite leaves arising from swollen nodes, regular flowers, usually with five petals, but sometime with the petals deeply notched or lobed to give the appearance of 10 petals per bloom, as seen in the chickweeds.

Chenopodiaceae—The Goosefoot Family (Dicot)

Common lambsquarters, whose young foliage is edible and as nutritious as spinach, chard, and beet (also members of the goosefoot family), hardly needs introduction to vegetable and grain farmers. Considered the world's tenth worst weed lambsquarters can be an aggressive spring or summer annual weed in croplands in temperate regions around the world. It is highly prolific, and one large plant can shed 50-100,000 viable seeds.

Young (a) and mature (b) lambsquarters plants.

Kochia is another summer annual weed in the goosefoot family. It has become a problem in annual crops in the northern Great Plains region.

Commelinaceae—The Spiderwort Family (Monocot)

Dayflowers are often mistaken for dicots because their leaves are wider than those of most monocots. They have three-petaled blue to purple flowers (in some species, one petal is smaller and paler than the other two), with six stamens. Asiatic dayflower and spreading dayflower are sprawling summer annual herbs that have become naturalized in moist soils near streams and bottomland woods throughout the eastern United States. They are occasional weeds of gardens and fields. A more recent and more invasive introduction is the Benghal dayflower, which spreads aggressively by rhizomes, and can form flowers and seeds either above or below ground. It has become a severe problem in cotton and other crops in parts of Florida and the Gulf Coast.

Asiatic dayflower (a) and Benghal dayflower (b).

Convolvulaceae—The Morning Glory Family (Dicot)

The morning glory family includes the incredibly deep-rooted (3 to 18 feet) wandering perennial field bindweed, recognized as the world's 12th most damaging weed the slightly less deep-rooted yet still highly invasive hedge bindweed and several summer annual morning glories that frequent vegetable and row crops in the South and Midwest. The parasitic dodders are also members of this plant family.

Figure shows hedge bindweed climbs this fence and covers the ground between clumps of grass weeds (a). Uprooted bindweed shows the vining habit and the underground network of brittle, white fleshy rhizomes (b). The one-inch fragment separated from the rest of the plant can regenerate a new plant from as deep as six inches. Ivyleaf morning glory (c) and tall morning glory (d) are two species of weedy annual morning glory that commonly occur in vegetable and row crops in warm-temperate climates.

Family characteristics include showy, funnel- or tube-shaped flowers, simple alternate leaves, and a climbing, vine-like habit of growth. Many of these weeds climb any upright support, including crop plants, whose growth and leaf opening can become seriously hindered as a result. Weeds in the Convolvulaceae thrive in summer heat, as does the sweet potato, one important food crop in this family. Sweet potato vines do not climb, but instead form a thick canopy, which effectively suppresses many weeds including yellow nutsedge.

Cyperaceae—The Sedge Family (Monocot)

While this family includes no major crops in the United States, and only a few weed species, one of the latter is purple nutsedge, once cited as the world's worst weed. This small, heat-loving plant causes serious economic losses in most tropical and warm-temperate crops around the world. A wandering perennial, purple nutsedge is highly allelopathic against many crops, and can form an underground biomass density of rhizomes, bulbs, and tubers that rivals or exceeds the total biomass of a vigorous 9-ft crop of sorghum.

Yellow nutsedge is also quite aggressive and has invaded colder regions up to the Canadian border. Researchers list yellow nutsedge as the world's 16th most costly weed. Cultivated strains of yellow nutsedge are grown in India for their edible tubers (chuffa).

Sedges are characterized by grass-like leaves deployed in a three-ranked arrangement, and solid stems distinctly triangular in cross section. In contrast, true grasses have rounded stems, which are hollow between the nodes, and leaves in a two-ranked arrangement.

Sedges are characterized by leaves arranged in three ranks around a solid, triangular stem (a), whereas true grasses have two-ranked leaf arrangement and hollow, cylindrical stems. This vigorous yellow nutsedge shoot (b) arises from a tuber in the top few inches of soil.

Euphorbiaceae—The Spurge Family (Dicot)

Cypress spurge is a major weed of vegetable and row crops in the north-eastern United States. Other common weeds in this family include leafy spurge (a major rangeland weed in the western United States), Virginia copperleaf, wooly croton, and spotted spurge. One of the characteristics of this plant family is the acrid, milky juice exuded from cut tissues, which can be a severe eye irritant. Flowers are small and inconspicuous, often imperfect (stamens and pistils borne on separate flowers), and often lacking petals. In many species, several flowers are grouped together in a cuplike structure, which itself resembles a flower.

Cypress spurge, Euphorbia cyparissias L.

Fabaceae or Leguminosae—The Legume Family (Dicot)

The legume family plays a special role in organic and sustainable agriculture, as these plants fix atmospheric nitrogen (N) through their symbiotic relationship with Rhizobium bacteria. Legumes provide protein-rich beans, peas, pulses, and livestock forages such as alfalfa, clovers, and lespedeza. This plant family includes a few summer annual weeds, such as hemp sesbania, and showy crotolaria, as well as kudzu, a highly invasive exotic perennial vine, which can overgrow and kill mature forest trees.

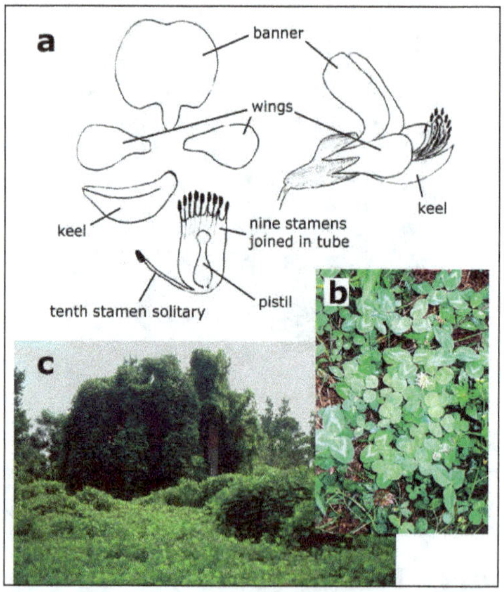

Many legumes have a characteristic pea flower (a), though some species such as mimosa have more regular flowers. Weeds in this family range from white clover (b) causing occasional problems in low-growing vegetables, to kudzu (c) smothering mature trees.

In addition to their N-fixing root nodules, legumes typically have compound leaves (trifoliate, pinnate, or palmate), and bear seeds in pods that dry and split open along both sutures. Many legumes have characteristic pea blossoms with petals differentiated into banner, wings, and keel; and 10 stamens with the filaments of 9 or all 10 fused into a strap or tube.

Lamiaceae or Labiatae—The Mint Family (Dicot)

The Lamiaceae includes several winter annual and perennial weeds, including purple deadnettle and ground ivy as well as many of our culinary herbs such as basil, oregano, sage, thyme, and mint. The mint family is characterized by aromatic foliage rich in essential oils, opposite leaves borne on stems that are square in cross section, and usually a distinctly bilabiate flower with four stamens. The flowers are rich in nectar and often attract and support natural enemies of insect pests.

Common weeds in the mint family include the winter annual purple deadnettle (a) and the perennial ground ivy (b).

Weedy members of this family that are familiar to vegetable and grain growers include henbit, deadnettles, and ground ivy. Mint itself can become a troublesome weed, as its perennial rhizomes readily spread from the mint patch into neighboring crops.

Malvaceae—The Mallow or Cotton Family (Dicot)

The mallow family includes the summer annual weeds velvetleaf, prickly sida, and spurred anoda, as well as the perennial common mallow. Velvetleaf is a major weed of corn and other row crops, especially in fields that are plowed annually. The mallow family also includes several important crops: cotton, okra, kenaf, hibiscus, and ornamental mallows.

The mallow family is characterized by showy flowers with five sepals, five petals, and numerous stamens whose filaments are fused into a tube surrounding the style of the pistil. Emerging seedlings of some species bear one roundish and one distinctly heart-shaped cotyledon.

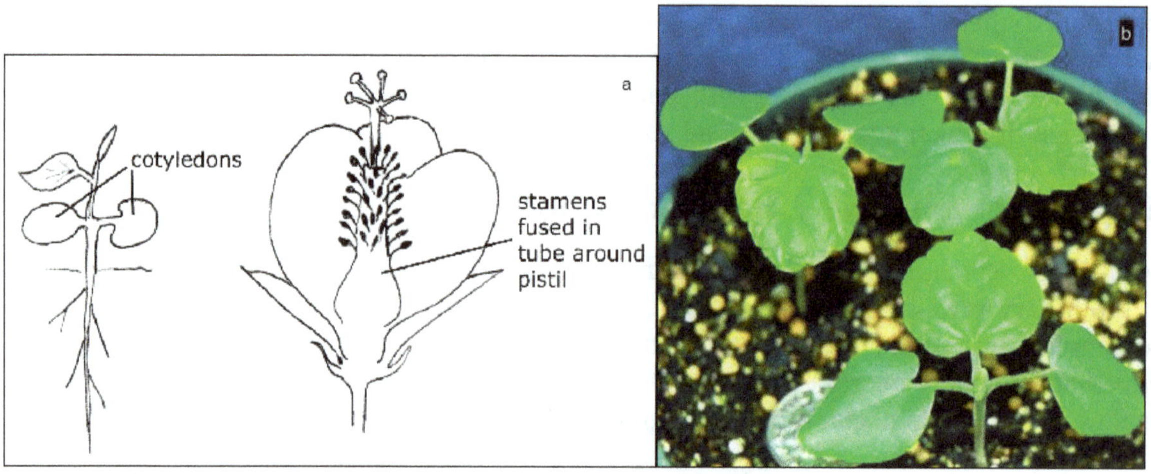

(a) Seedlings of some species in the mallow family have one roundish and one heart-shaped cotyledon. The flower has numerous stamens fused into a tube surrounding the pistil.

Oxalidaceae—The Oxalis Family (Dicot)

Yellow woodsorrel is a small rhizomatous perennial weed with yellow flowers. While it is most often a weed of turf, landscapes, and greenhouse plants, it can occur in vegetable fields, especially under reduced tillage. It has trifoliate compound leaves, and small, regular flowers with five yellow petals.

Yellow woodsorrel, Oxalis stricta L.—also known as common oxalis.

Phytolaccaceae—The Pokeweed Family (Dicot)

Pokeweed is a well-known and spectacular herbaceous perennial. It comes up from heavy fleshy taproots every spring and can grow as tall as 10 feet. It bears small white or pink flowers on racemes up to 12 inches long, and forms black berries 1/4 to 1/2 inch across. The foliage and berries are poisonous to humans and livestock, though newly emerging foliage can be eaten if properly prepared.

American pokeweed, Phytolacca americana L.

Plantaginaceae—The Plantain Family (Dicot)

The plantain family includes several weedy species, such as broadleaf plantain and buckhorn plantain, which are most troublesome in turfgrass, though they can occur in cultivated fields and gardens as well. The weedy plantains are low-growing perennials characterized by parallel leaf veins (unusual in dicots). The sessile (without petiole) leaves grow in a basal rosette, from which arises a leafless stalk bearing a dense spike of tiny, inconspicuously colored flowers.

Broadleaf or common plantain, Plantago major L.

This plant family also includes plants that were formerly classified in the Scrophulariaceae or Figwort family, such as yellow toadflax common mullein and several species of speedwell that occasionally occur in vegetable crops, as well as ornamental flowers such as snapdragon.

(a) Yellow toadflax, Linaria vulgaris P. Mill. (b) Common mullein, Verbascum thapsus L.

Poaceae or Gramineae—The Grass Family (Monocot)

The grasses comprise a large family of monocots—clearly the most agriculturally important plant family on Earth. In addition to many of the world's staple food and fodder crops (corn, rice, wheat, millets, sorghums, other annual cereal grains; and many perennial and annual forages), the grass family contributes 29 of the 76 species listed as the most important agricultural weeds worldwide, including 10 of the top 18. Bermuda grass, barnyard grass, jungle rice, goosegrass, johnsongrass, and large crabgrass—which are all too familiar to farmers in the United States—take second, third, fourth, fifth, sixth, and eleventh places in this listing. The weedy species include both annuals (e.g., barnyard grass, goosegrass, crabgrasses, foxtails) and perennials (e.g., Bermuda grass, johnsongrass, quackgrass).

Two major weeds in the grass family include goosegrass (a) and johnsongrass (b).

The most definitive characteristics of grasses are that the single cotyledon remains underground inside the seed after germination, and that the seedling shoot emerges as a rolled or folded set of leaf blades protected by a sheath (coleoptile) that opens at the soil surface to release the expanding leaf blades. The growing point remains underground for some weeks after emergence unlike broadleaf plants thus the young grass plant is not killed by frost, heat, mowing, or removal of aboveground parts. As a result, young annual grass weeds are more difficult to control by cultivation or flame weeding than annual broadleaf weeds of the same size. Mulching is also less effective on grasses than other weed seedlings, as the coleoptiles can penetrate organic mulches more readily than the delicate shoots of broadleaf seedlings.

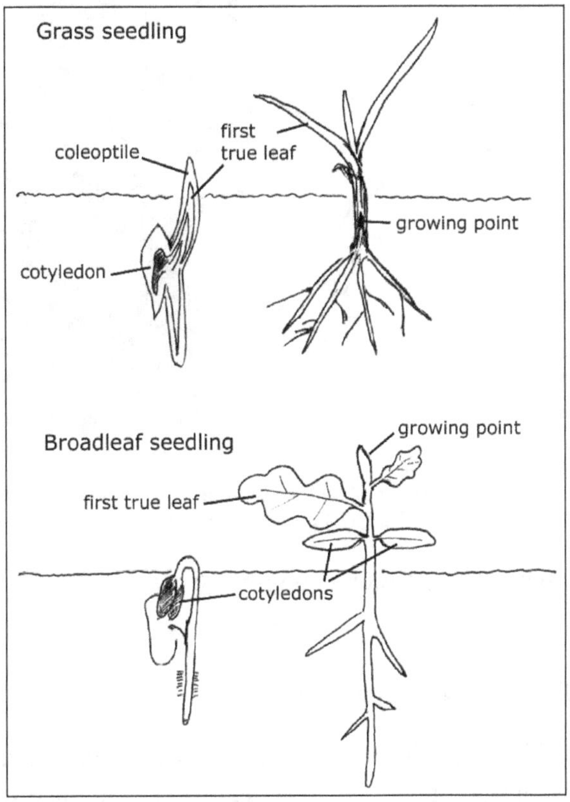

Figure shows Grass seedlings emerge as a pointed coleoptile, which protects the new leaves as the seedling pushes through the soil and surface residues. The growing point remains below ground for several weeks and can regenerate the plant if the top is severed or frozen. Most broadleaf seedlings bring both growing point and cotyledons above the soil surface upon germination, leaving them more vulnerable to cultivation or unfavorable weather conditions such as frost.

Other grass characteristics include narrow, linear leaves (often 20-50 times as long as wide); small, inconspicuous flowers borne on a spike or panicle at the top of the stem, small dry fruits each containing a single seed or grain, and a strong, fibrous root system. Perennial grasses may or may not have stolons or rhizomes, but never taproots. Quack grass, Bermuda grass, and johnsongrass are all wandering (creeping) perennials that propagate through strong, aggressive rhizomes or stolons.

Polygonaceae—The Buckwheat Family (Dicot)

The buckwheat family includes the knotweeds, smartweeds, curly dock, broadleaf dock, and red sorrel, as well as buckwheat, a valuable grain and cover crop. Wild buckwheat and mile-a-minute are rampant annual vines that climb, shade, and bind other plants much like bindweeds and morning glories.

One distinguishing characteristic of the buckwheat family is the ocrea, a prominent membranous or leafy sheath encircling the stem and base of the petiole at each node. Other traits include small flowers, which may be white, greenish, or brightly colored; simple leaves with mostly smooth (entire) edges, and a small, dry, single-seeded fruit, which is often three-sided or angular.

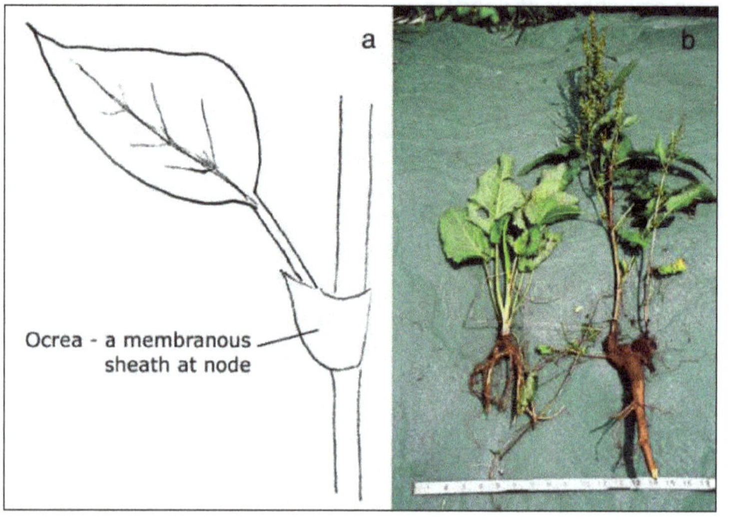

Plants in the buckwheat family have a characteristic membranous sheath around each leaf node (a). Broadleaf dock and curly dock (b) are taprooted perennial weeds in this family.

Some species, notably buckwheat and smartweeds, have plentiful nectar that can attract and support beneficial insects.

Portulacaceae—The Purslane Family (Dicot)

Common purslane is an edible, succulent summer annual weed, which grows in prostrate mats and can compete severely with young or low-growing crops. It is hard to control by cultivation, as broken fragments of this drought-tolerant weed re-root readily and grow into new plants. These traits, combined with its worldwide distribution, have earned it the dubious honor of being the world's ninth worst weed. Cultivated members of this family include ornamental flowering portulacas as well as some domesticated varieties of purslane grown as a salad green.

Common purslane is edible and highly nutritious. It does not compete aggressively when it grows as an understory in tall, established crops such as sweet corn or tomato, but it can be a major problem during crop seedling establishment and in slow-growing vegetables like carrot or onion.

Rosaceae—The Rose Family (Dicot)

The rose family includes many of our most important fruit crops—strawberries, brambles, apples, pears, peaches, and other stone fruit—as well as almonds, cultivated roses, and some other ornamental crops. Multiflora rose escaped from early colonial gardens to become a serious pasture weed. Wild blackberries and other wild brambles are sometimes considered weeds, and sometimes welcomed as a wild food source. Vegetable growers encounter multiflora and brambles mainly when they first convert pasture, hayfield, or brushy land to vegetable production. The cinquefoils are low-growing herbaceous perennial weeds in this plant family, which are usually only a minor problem in cropland.

Characteristics of the rose family include alternate leaves, regular flowers with five sepals, five petals, and many stamens in several whorls. (Cultivated roses have been bred/selected to have many petals per flower.) The family includes perennial trees, shrubs, and herbs, many of which are armed with thorns or prickles.

Rubiaceae—The Madder Family

This plant family includes one major weed of winter grain crops: cleavers, otherwise known as catchweed bedstraw. It is not considered a major weed in vegetable production.

Cleavers, Galium aparine L.

Solanaceae—The Nightshade or Tomato Family (Dicot)

The nightshade family includes several toxic weeds such as the summer annuals jimsonweed and black nightshade, and the rhizomatous perennial weed horsenettle which likes to leave its tiny, irritating spines in farmworkers' hands. Major food crops in this family include potato, tomato, sweet and hot peppers, and eggplant. In addition to competing for nutrients and moisture, nightshade weeds in vegetable fields can be a reservoir of pests or diseases of these crops.

Horsenettle, shown here in pasture, can also occur in vegetable fields, especially those recently converted from sod or managed with reduced tillage.

The nightshade family is characterized by regular (symmetrical) flowers with five petals partially or fully fused into a tube or bell-shaped corolla, fleshy fruits (berries), and usually simple leaves. In most species, the foliage accumulates toxic alkaloids and can be dangerous if ingested by livestock or humans. If the dark, fleshy fruits (berries) of black nightshades contaminate vegetable seed (pea or bean) crops during harvest, they can stain the seeds and render them unmarketable.

Classification According to Nature of Stem

Woody Weeds

These hardy plants come back year after year and are as tough as the woody stems that set them apart from other lawn weeds. There is a way to stop them, but it takes consistency especially if you don't want them coming back.

What is a Woody Weed?

Vines, thorny branches, thick stems — if there's a weed in your yard that looks like it belongs in a wooded area, it's more likely a woody weed. Even if you don't live near a wooded area, the seeds from these weeds show up in your yard if transported by wind, rain, mulch, and birds. Some examples include honeysuckle, poison ivy, clematis vine, wisteria, and English ivy. Like most weeds, left alone these plants quickly overtake a lawn or garden.

Weed Control

The easiest way to help prevent any weeds from taking over is regular lawn maintenance. Keep the weeds from getting established by taking action as soon as you see one pop up. Here are a few suggestions:

- Use mulch, landscape cloth, or other physical barriers throughout your garden to prevent weed seeds from germinating.

- Dig up any invasive plant roots and discard it. Pulling it by hand can stimulate the plant to grow more and can also be dangerous if it's poisonous or thorny. You have to get the whole root to ensure it won't return. If the plants are too big to dig out, cut the stem close to the ground, and apply an herbicide directly on what remains and cover with dirt.
- Leaves and larger woody weeded areas can also use an application of herbicide. Just be careful not to treat any plants you want to keep.
- Time the application of herbicides when the plant is more susceptible – usually when it flowers. This varies depending on the plant.

Any time you use a chemical treatment, it's best to follow the directions closely for a more safe and effective process.

Classification According to Association

Crop Weeds

Crop weeds are weeds that grow amongst crops. Despite the potential for some crop weeds to be used as a food source, many can also prove harmful to crops, both directly and indirectly. Crop weeds can inhibit the growth of crops, contaminate harvested crops and often spread rapidly. They can also host crop pests such as aphids, fungal rots and viruses. Cost increases and yield losses occur as a result. Striga, one of the main cereal crop weeds in Sub-Saharan Africa, commonly causes yield losses of 40–100% and accounts for around $7 billion in losses annually. Around 100 million hectares of land in Sub-Saharan Africa are affected by striga. Barnyard grass has been identified as a culprit in global rice yield losses and certain species have been known to mimic rice. Examples of crop weeds include chickweed, barnyard grass, dandelion, striga and Japanese knotweed.

Grassland Weeds

The principles in eradicating weeds from grassland are those which make for general improvement in the herbage, and in many cases attention to drainage, manuring, grazing, etc., rather than direct methods of destruction (though spraying is considered), will lead to the eradication of undesirable species. A large number of the worst weeds that occur on grassland are dealt with individually, classification being made according to the natural orders to which they belong).

Docks (Broadleaved or Curled)

One plant can produce 60,000 seeds that are viable for up to 80 years. Open swards as a result of poaching, over-grazing or winter kill provide space for infestations to start. Docks thrive in fertile pasture but only provide 65% of the feed value of grass from the same area. Docks are best controlled at the rosette stage when leaves are healthy and not under stress.

Thistles (Creeping or Spear)

Creeping Thistle is a perennial weed, growing from seed or root. Spear Thistle is a biennial growing from seed, often unnoticed in the first year due to a small rosette. Year two it can spread vigorously. Established Creeping Thistle has extensive underground roots and competes strongly with grass.

Spear Thistle in the second year can spread to cover more than a square metre of ground seriously reducing pasture productivity. Thistles can appear at different times thus timing of control can be difficult. Treatment following topping ensures a lasting effect.

Nettles

A perennial weed that can grow from seed or rhizome. As Nettle infestations grow they spread making pasture unpalatable reducing grazing area. Poached and open swards will encourage Nettles to establish. Once established, Nettles are rarely controlled by one application and require a programme of treatments.

Ragwort

Ragwort can be poisonous to stock. Overgrazed or bare ground will encourage Ragwort establishment. Uprooting Ragwort will not be effective as the rhizome will break off and re-grow. Ragwort is best controlled at the rosette stage.

Common Chickweed

Rapid prostrate growth means it competes aggressively with grass, leading to significant losses of yield especially when establishing new leys with up to 25% reduction in silage yield. Chickweed is a problem in re-seeds.

Creeping Buttercup

A common weed particularly in new or worn-out pastures. Buttercup is mildly toxic and stock will not graze it. The weed will thrive in overgrazed land, especially where drainage and fertility is poor.

Soft Rush

A common weed on wet pastures. Rushes colonise rapidly over extensive areas and reduce productivity. Poor drainage and acid soils encourage Rush growth. Topping Soft Rush in late spring prior to seed formation weakens the plant, which then has to re-grow. Treat re-growth when the Soft Rush is about 15cm tall for best control.

Common Grassland Weeds

Five key reasons to control broadleaved weeds in grassland:

- Maximise Grass Yield: As a general rule, 1% ground cover of grassland weeds, such as Thistles and Docks, will reduce the productivity of grassland by 1%. If grassland weeds are as high as 10% then severe yield losses will occur and this will happen year on year unless a programme of weed control is planned.
- Make Better Quality Silage: Weed infestations affect the quality of silage particularly Docks and Thistles.

- Better Grassland Utilisation: Grazing animals tend to avoid patches of weeds, especially thistles. As a result there is a ungrazed grass around the weed so a decrease in pasture utilisation.
- Healthier Stock: Weeds pose a threat to livestock health. Injurious weeds such as Ragwort can result in illness and even death, particularly in cattle. Thistles can act as "hypodermic needles" spreading diseases such as Orf in grazing sheep and lambs.
- Extend Pasture Life: Broadleaved weeds allow less productive grasses to invade, particularly after cutting. This results in the reduction of desirable grass species and earlier renewal of the pasture is required.

Aquatic Weeds

It is unanimously accepted that the main concern of the commercial fish farm is the profitability. One of the prime difficulties is uncontrolled growth of aquatic weeds. The control of aquatic weeds in almost all the systems of aquaculture is a problem noticed all over the world. The weeds problems are more in aquatic systems of tropical and semitropical countries like India and then in the western countries. The weeds can be defined as unwanted and undesirable plants that are adopted to grow and reproduce under aquatic conditions.

Limited growth of aquatic plants may be useful in the maintenance of water quality and may serve as shelter and substrate for food organisms in water but their uncontrolled growth may cover the surface of the water of the pond resulting large numbers of problems, which are mentioned below:

- The penetration of the light will not be adequate and thus affect the productivity of fish production.
- The nutrients in the water, which is generated by fertilizing the pond, will be consumed more by plants, and fish will not get the sufficient nutrients, hence it will badly affect the growth and production rate of the cultured species.
- Blooms of the algae often results in the depletion of oxygen (as a result of dead and decaying algal mass that quantity of oxygen production will be reduced) causing anoxia (less quantity of oxygen) for the fishes cultured in the pond.
- Dissolved oxygen (DO) is the parameter for water quality and reflects the physical and biological processes prevailing in the water. To maintain good fish population in the fishpond, the DO should be around 6.0 mg/lit.
- Fishing with nets will be difficult. It will be a problem for cage fisheries also as the algae will stick to the cages.

Common aquatic weeds comprise both filamentous and single cells algae:

- Floating Weeds: They are free floating with their leaves over the surface of water while roots are under water. The most common are Eichornia, Azolla and Pistia.

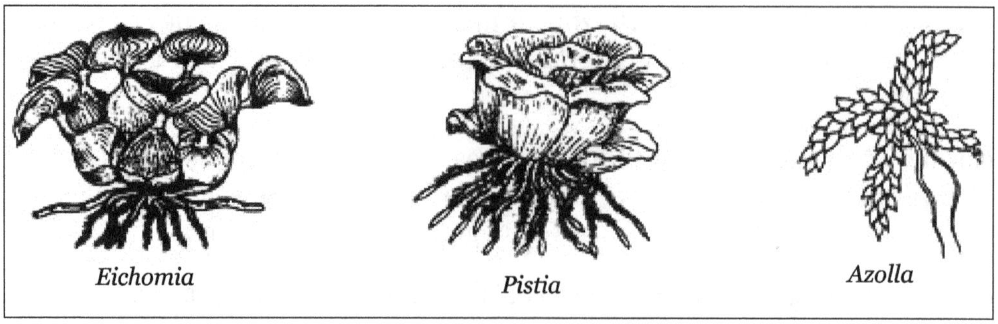

Floating Plants.

- Emergent Weeds: They possess roots which are present in the soil of the water while their leaves or shoots are present over the surface of water, e.g. Utricularia Nymphaea, Trapa, Myriophyllum, Otella, Vallisneria etc.

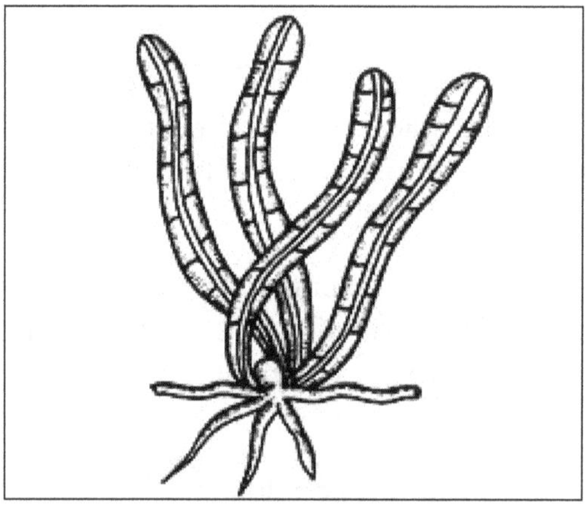

Anchared Submerged Hydrophytes.

- Submerged weeds are those weeds, which are completely submerged under water, but their roots are present in the soil of the pond. The common examples are Hydrilla and Naja. The Ceratophyllum and Utricularia also come under the category of submerged weeds but their roots are not present in the soil and the roots are free floating.

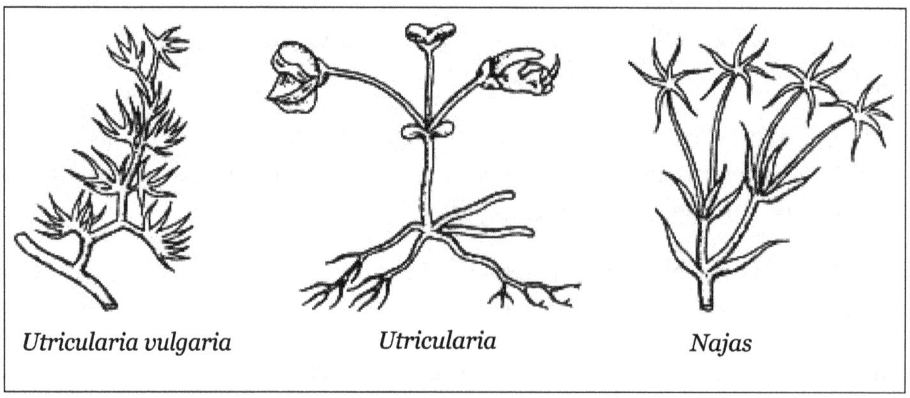

Submerged Weeds.

- Marginal weeds grow on the sides of the pond and are rooted in the waterlogged soil. The common marginal weeds are Typha and Phargmites.

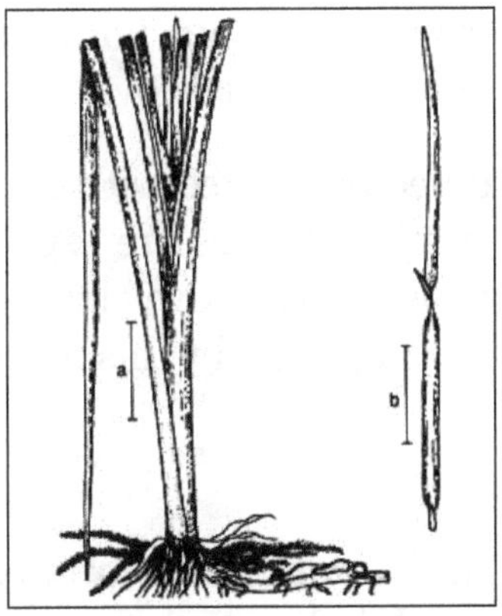

Typha dominengesis (a) sterile shoot (4 cm) (b) inflorescene (8 cm).

- Filamentous algae, form mats in the marginal area or scum in the main water body. The most common filamentous algae in the ponds are Spirogyra and Pithophora.
- Planktonic algae proliferate very rapidly and form algal bloom in water. The most common example is Microcystis and Anabaena.

Control Management of Aquatic Weeds

Following are the methods for the control of weed:

- Manual,
- Mechanical,
- Chemical,
- Biological,
- Control by intra-specific competition.

Manual Method

The manual method is the most convenient, cheap and easy and in the developing nations, since there the labour is cheap. It is not long lasting because the weeds grow again as it is difficult to eradicate the weeds completely.

The emergent and marginal weeds are removed by pulling them with hand. They are kept under check by cutting their floating roots repeatedly. The floating weeds are removed either by hand or wire-coirs nylon-nets.

For the clearing of rooted submerged weeds simple methods like hand-pull bottom rakes, bamboo poles with toothed pronged can be used. Repeated removal, combined with biological or chemical method, may be continued to control the weed.

Mechanical Methods

The mechanical methods are many; their application depends upon different size of lakes. They are different if they are used for extensive fish farms. For moderate fish farm of 100 ha, the most common method is the weed cutter. It is used for removal of submerged and emergent weeds.

The weed cutters are either flat-bottomed boat fitted with cutting beams or other cutting devices. Amphibious boats fitted with weed-cutting devices are especially convenient for use in shallow ponds and enclosures. It may appear easier to remove algal growths such as musk grass (Charra) but it is difficult to remove the clumps completely by this mechanical method.

Chemical Methods

The following is the list of chemicals used for the eradication of the weed. The use of these chemicals has adverse effect on the fish growth indirectly because these chemicals badly affect aquatic organisms which are food of the fish. Some of the herbicide and weedicide, chemical used are Sodium arsenite, xylene, 2, 4 dichlorophenoxy acetic acid (2, 4 D), etc.

Biological Control of Aquatic Weed

There are large numbers of biological control methods. The use of herbivorous fishes and other aquatic animals has been used for biological control of weeds. The biological control of aquatic weeds is easy and popular and is achieved by fishes those are phytophagus or herbivorous fishes.

The most common fishes are Ctenopharyngodon idella (grass carp) Carassius carassius, Carassius auratus, Tilapia mossambicus. These fishes consume aquatic weeds as their food. These fishes are having strong pharyngeal teeth and, therefore, could macerate plant material.

Hypophthalmichthys molitrix feeds on phytoplankton. It is reported that grass carp could consume about 19.9 metric tons of water weed and could produce about 195 kg of fish flash. It can thus control growth of aquatic weed very efficiently. The aquatic weed can be used as feed for fishes. It can also be used as protein for human consumption. The following parts of the weeds are used as food for human use:

- Lpomea aquatica, young leaves and stem, leaves.
- Marsilea species, flowers, leaves and rhizome.
- Nymphia species, seeds.
- Euryale ferox are used for human consumption.
- Weeds are also used as the source of energy and biogas.

The common weeds are water hyacinth. The cow dung and water hyacinth are mixed in the ratio of 1:1.

Waste Water Treatment

Some aquatic plants such as Scirpus lacustris, Ceratophyllum demersum, Spirodella polyrhiza and Lamina minor are used in waste-water treatment. Some aquatic weeds are used in pulp, paper and fibre industry. There are some other uses of weeds. They are used for the construction of huts, etc.

Forest Weeds

When defining a weed, it should be emphasized that there is no simple and precise definition. Weeds are plants growing among cultivated plants and interfere with man's activities; weeds are plants growing in places they are not wanted. The concept of a weed is relative in nature, i.e., there is no sharp boundary between weeds and cultivated plants. While in some regions, one plant species is considered a weed, the same species is cultivated in others.

Vajda considered as forest weeds those plant species interfering with germination and growth of young forest plants and Konstantinovic those which were unfavourable under certain circumstances in the forest and interfere with forest management. The weeds in forestry are all herbaceous plants, shrubs, and trees which, in forest nurseries, stands, and clear felled areas weaken or prevent the growth and development of cultivated trees.

Evidence of the benefits of weed control for enhanced tree growth is widespread however weeds in forestry are not always harmful. Herbaceous weeds in forest plantations represent food for livestock and dry weights of some weeds are used as bedding for livestock. The fruits of some weeds are edible and weeds while they are somewhere used for human consumption. Some weeds have medicinal properties and are used as medicinal plants. Weeds prevent soil erosion and can be a shelter for wild animals and birds. However it should be noted that the benefits of weeds significantly less than the damage caused.

Damage from Weeds

Weeds present a large challenge both in agriculture and in forestry. They form a large mass of aboveground and inground organs engaged in a competitive relationship with cultivated plants for light, water and nutritive components in the soil. Damages caused to cultivated plants by weeds can be great. According to some opinions, damages caused by weeds are greater than those caused by diseases and pests together. Weedy plants grow relatively faster in forest settings and displace young forest plants living space, overshadow and stifle them, and water and nutritive matters are taken at the expense of cultivated plants.

Far less favorable impacts of weeds are found in nursery production. Due to the presence of weeds, nursery plants can experience retardation of growth, chlorosis, reduced resistance to plant diseases and pests, and death of individual parts of branches or crowns; if the weed is abundant, it often leads to drying and deterioration of the entire plant. If weed control in nurseries is lacking, nursery plants of poor quality and fewer total plants are obtained.

Distribution of Weeds in Forestry

Competitive division of weeds in forestry is often made according to the degree of harmfulness of weeds to the trees. Weeds in forestry are classified as either useful or harmful. We can classify weeds into useful, harmful, or indifferent. According to this classification:

- Harmful weeds are plants that hinder tree development, and form thick cover.
- Indifferent weeds are plants that grow individually, form weak coverage and do not hinder development of cultivated plants.

- Useful weeds are plants with medical properties and plants that form fruits.

The role of light has been of particular importance for emergence of weeds. In relation to light regime, weeds may be classified into sciophytes – plants developing in the shadow in weakly thinned forest stands or in dense forest stands and represent no threat to tree development; semisciophytes – semi-shadow plants that develop in thinned stands and can do a lot of harm; and heliophytes – plants of open habitats such as clearings, strips, burnt areas, etc., and represent a big threat to renovation and development of trees. There are a number of other weed classifications due to their adaptation to abiotic factors such as water regime, temperature, physico-chemical soil characteristics, etc. during their evolutionary development. However, very important weed classifications in forestry, which would have practical significance from the aspect of weed control, are the following weeds of forest nurseries and weeds of forest plantations and forest stands.

The most Important Weeds in Forestry

Weeds in Forest Nurseries

Weed flora in forest nurseries differ from those found in forest plantations and forest stands. Given the extent of care measures applied, weeds in forest nurseries are very similar to those found in cultivated crops. They are mostly annual and perennial herbaceous weedy species. The most common grass weed species present in the forest nurseries include: Sorghum halepense, Cynodon dactylon, Alopecurus myosuroides, Digitaria sanquinalis, Echinochloa crus-galli, Poa annua, and Setaria spp. Dominant broadleaf species include: Amaranthus retroflexus, Ambrosia artemisiifolia, Chenopodium album, Cirsium arvense, Convolvulus arvensis, Erigeron canadensis, Datura stramonium, Galium aparine, Solanum nigrum, Sinapis arvensis, and Poligonum spp.

Control of weediness in forest nurseries is very important and quality planting material is the basic prerequisite for success in forest stand establishment. Since weeds are one of the most limiting factors for the success of nursery production, their control should be approached very seriously.

Weeds in Forest Plantations and Forest Stands

Weeds in forest plantations and forest stands differ from those in forest nurseries, because, in addition to different care measures applied in plantations and stands, the conditions in habitats also differ. Apart from ferns, herbaceous annual and perennial weeds, woody weeds such as shrubs, bushes, and shoots from the stumps of different tree types may also be present in forest plantations and stands. Woody weeds are very hardy and have a great power of regeneration; it is practically impossible to destroy them completely by mechanical means. The most common weed species present in forest plantations and stands are: Ambrosia artemisiifolia, Amorpha fruticosa, Asclepias syriaca, Erigeron canadensis, Solidago gigantea, Sorghum halepense, Sambucus nigra, Stenactis annua, Pteridium aquilinum, Rubus caesius and etc.

Weed Control in Forestry

There are numerous measures and procedures for weed control in forestry today, but, in order to fight weeds successfully, they should consist of different care and control measures. Described below are the six classifications of weed control measures.

Preventive Measures

The main goal of preventive measures is to prevent weed distribution. All measures used to protect any surface from weeds, i.e. to prevent weed seed growth in the field are considered preventive measures. Preventive measures in forestry weed control include:

- Control aided by sowing only pure crop seeds, which prevents spreading of weeds over sown surfaces.
- Destruction of weeds on non-agricultural areas; weeds that present a constant source of weediness and transportation of seeds to arable lands are developed on such areas.
- Prevention of the spread of weed seed by human activities by keeping agricultural and forest machinery and objects clean.
- Allelopathy is a phenomenon, where cultivated plants secrete exudates affecting the suppression of weeds. It is manifested in such a way that in the presence of certain plant species, many others are not able to thrive, or are slowly developed.

Mechanical Measures

Mechanical measures for combating weeds include basic treatment such as ploughing, disking, tilling and etc. Also regular measures in forest nurseries and plantations are hoeing and farrowing, undertaken during the greatest part of the vegetation period and especially emphasised during the entire spring and in early summer.

One of the ways of suppressing the already growing weeds and preventing their seed dispersal is mowing. Multiple repetitions exhaust the stored substances in the root and the plant is killed. In addition to mowing, one of the methods of weed suppression in forestry is also the pruning of shoots and stump shoots. However, this weed suppression method is relatively expensive due to intense labor and if repeated pruning is required depending on the weed species present. Concerns about increasing pesticide use have been major factors for research in physical weed control methods.

Physical Measures

Physical weed control measures applied in forestry involve the use of flame and superheated steam. Destruction of weeds by flame can be applied in forest plantations with wider spaces between the rows, provided that the crops are previously protected by metal shields. Burning weeds is carried out on non-productive areas such as forest railways, roads, and canals. Destruction of weeds using steam is applied in forest nurseries in preparation of substrates used for sowing or planting. This is also a form of sterilization which destroys weed seeds in addition to plant diseases and noxious insects. The use of flame for the reduction of weeds is the best at the end of growing seasons, because in this way destroy most weed seeds that are dispersed on the soil surface.

Mulches

The covering of soil with a variety of materials such as straw, stubble, polyethyelene films, and others, to prevent the emergence of weeds is utilzed on smaller areas, mostly in forest nurseries. Polyethylene films of varying colors and thickness are most often used. This type of weed control is efficient for annual weeds but has no effect on control of many perennial weeds, and can be expensive compared to other methods used to fight weeds.

Many types of mulches have been tried including: sheets of plastic, newspaper, plywood, various thicknesses of bark, sawdust, sand, straw, sprayed-on petroleum resin, and even large plastic buckets. Most have proven to be ineffective, costly or both. Early trials tended to use small, short-lived materials that aided conifer seedling survival but not growth. Compared to other weed control techniques available in previous years, mulches were rather expensive. Current trends are to apply longer-lived, somewhat larger mulches of mostly sheet materials made of reinforced paper, polyester, or polypropylene.

Biological Weed Control

Biological measures of weed control are based on the application of natural weed enemies such as insects, fungi, viruses, and bacteria in order to prevent their dissemination, and thus spreading. There are numerous examples of successful biological weed control. Application of pathogenic fungus, Chondrostereum purpureum, is used to control beech, yellow birch, red maple, sugar maple, trembling aspen, paper birch, and pin cherry. Exotic leaf pathogens, Phaeoramularia sp. and Entyloma ageratinae, were used for control of Ageratina adenophora and Ageratina riparia in South Africa. Control of Hypericum perforatum can be done by using insects Chrysolina quadrigemina and Zeuxidiplosis giardi. For control of Acacia longifolia, the widely spread invasive plant species in Portugal, the bee wasp Trichilogaster acaciaelongifoliae was used. In those parts of the world where Eucalyptus sp. presents a problem the pathogen, Cryphonectria eucalypti, may be used for its suppression.

Application of biological measures in weed suppression has its limitations, though it has several advantages. Cultivated plants can be protected from some weeds, but not from all of them. It is impossible to destroy weeds completely because the biological agent depends upon the weed for survival; moreover, it is difficult to program biological protection for numerous cultivated plants from weeds with certainty since there are many similarities between weed species and cultivated plants.

Herbicides

Herbicides are used in forestry to manage tree-species composition, reduce competition from shrubs and herbaceous vegetation, manipulate wildlife habitat, and control invasive exotics. Unlike agriculture, the use of herbicides in forestry began much later and generally the application of herbicides in forestry was based on experiences from intensive agricultural production. The results of research in agriculture are applied in forestry with major or minor delays. Due to the lack of labour, high labour costs, and large areas, producers are more often interestedin the use of herbicides. Use of herbicides in forestry decreases weediness, particularly at the initial stages of development of forest nursery plants, when the effect of weeds on plants is the greatest; at the same time, much better economic efficiency in the production process is achieved. Also, possible mechanical damages to the nursery plants can be avoided, and it happens very often that any kind of mechanical treatment is prevented in early stages of plant development due to high soil humidity. Use of herbicides to control competing vegetation in young forests can increase wood volume yields by 50– 150%.

Division of Herbicides

According to the type of action herbicides may be divided into the herbicides with contact action and herbicides of translocation. Contact herbicides destroy above ground parts of plant, only the parts of the plant they touch. Translocation or systemic herbicides absorbed by leaves are transferred through the whole plant.

According to the mode of action herbicides may be divided into the total and selective herbicides. Total herbicides kill all plants and selective herbicides kill weeds, and are not harmful to cultivated plants. According to the time of application herbicides may be divided into the herbicides applied before sowing or planting, herbicides applied after sowing, and before emergence of weeds and cultivated plants and herbicides applied after emergence of weeds and cultivated plants.

Mechanisms of Action Herbicides

For contact herbicides, action is manifested at the site of penetration. Contact herbicides penetrate quickly through cuticle and epidermal cells of plants and the toxic effects on weeds are quickly observed. In systemic herbicides, absorption may occur through the root, stem, and leaf. If an herbicide is absorbed through the roots it can move via the xylem to the above ground parts; more often, it moves to leaves where disturbed respiration and photosynthesis occur. Herbicides absorbed by the leaves and stem cause harmful effects in absorbed plant parts and by spreading via the phloem to reach the root and ground plant organs (rhizomes). Whether the plant will absorb a higher quantity of the herbicide through above ground parts or the root system depends on herbicide application, type of herbicide, and several other factors.

Herbicides exhibit different mechanisms of action. Some herbicides inhibit synthesis of amino acids in plants, and others help the formation of free radicals in plants. Lipid synthesis is the site of herbicide action used to control monocot weeds, and a great number of different herbicides inhibit the process of photosynthesis. While some herbicides act only on one process, others act on multiple processes in plants. If only one process in the cell is disturbed, the whole range of processes is affected. Due to that it is difficult to determine the primary herbicide action and consequences.

Selectivity of Herbicides

Selectivity is a property of herbicides to destroy weeds effectively without harming cultivated plants. Selectivity is not an absolute property of any herbicide. There is a whole range of factors such as morphological, biological and physiological plant traits, chemical composition, and herbicide structure, quantity, mode and time of herbicide application and translocation of herbicides on which selectivity of some herbicides depend. Herbicide selectivity is a dynamic process with complex interactions between plant, herbicide and environment.

The main mechanism of herbicide selectivity is the differential metabolism between weeds and crop species, by which susceptible weeds are less able to metabolize selective herbicides. The importance of understanding the main stages of differential metabolism- based selectivity derives from the elucidation that plants, to an extent, use cell energy to process and detoxify herbicides.

There is also morphological selectivity based on plant structure. Leaf structure such as vertical and narrow leaf, waxy coat, and a protected vegetation cone contribute to the fact that some grasses are resistant to 2, 4-D herbicide. Greater resistance of coniferous species to herbicides compared to broadleaves is also based on leaf structure.

Degradation of Herbicides

A great part of the total quantity of herbicides applied in agriculture and in forestry is found in the soil. After introduction into soils, several processes affect the vertical and horizontal distribution of the herbicides including transport by water flow, sorption to soil components and various degradation processes. Degradation can involve biotic and abiotic processes, where microbially facilitated biodegradation is especially interesting, as it is a major process in the complete mineralisation of compounds to harmless inorganic products.

Decomposition of herbicides in the soil is a complex process taking place in several stages such as photodegradation, chemical degradation, and microbiological degradation. Photodegradation is means that some herbicide molecules such as trifluralin, dinitroaniline are degraded by the influence of ultraviolet rays, and these herbicides should be incorporated into soil after application. Chemical degradation of herbicides in the soil is done through processes of oxidation, hydrolisis, hidratation and reduction during which the herbicides are completely or partly degraded. Microbial degradation plays an important role in herbicide breakdown, and of their toxic material found in soil. Herbicides not only influence the activity of microorganisms, but the fate of herbicide in the soil depends on the activity of microorganisms. Abilty of microorganisms to carry out herbicide biodegradation depends on the type of applied herbicide, herbicide chemical properties, applied herbicide concentrations, and great number of biotic and abiotic factors.

The Toxicity of Herbicides

Toxicity is the capacity of a substance to harm or disturb the health of an organism. Toxic effects may be immediate (acute) or accumulative (chronic), depending upon the exposure duration, the dose, and the herbicide. The toxicity of a substance varies with the animal species, age, sex, and nutritional status and with the route of exposure—through the stomach (orally), the lungs (by inhalation), or the skin (dermally). The skin and eyes are also subject to irritation caused by chemicals.

A common way to document toxicity is by oral LD50 values. LD50 is the amount of chemical required to provide a “lethal dose” to 50% of the test population. LD50 is measured in mg of chemical administered per kg of body weight. Toxicity tests are conducted on experimental animals, such as white rats, mice, and rabbits. Due to different ways of herbicide action when estimating its toxicity the whole range of data should be known and therefore it is difficult to express herbicide toxicity.

Impact of Herbicides on the Environment

The impact of herbicides on the environment (water, soil, biodiversity, etc.) may have diverse effects depending on the whole range of factors and, initially, on the environment in which it is found after application. In general, herbicides most commonly used for vegetation management in forestry (glyphosate, triclopyr, imazapyr, sulfometuron and etc.) degrade quickly once they enter the environment and thus are neither persistent nor bio-accumulative. Forest herbicides persist short term in the environment, and have few toxic effects when operationally applied following herbicide labels. Single applications of forestry herbicides at stand initiation have minor and temporary impacts on plant communities and wildlife habitat conditions. Studies carried out on the effect of herbicides hexazinone fosamine ammonium and glyphosate in forestry have revealed that these herbicides have minimal effects on soil microorganisms and exhibit little or no potential for bioaccumulation. If herbicides are properly used, current research indicates that the negative effects on wildlife usually are short-term and that herbicides can be used to meet wildlife habitat objectives.

Orchard Weeds

In orchards some forestry plants like shahtoot; peepal and tahli are often seen growing. This is so because after fruit harvest and during juvenile period, birds rest on fruit plants without any disturbance. The tree weed seeds are often defecated under fruit tree. Tree plants like peepal and mulberry grow under the canopy or near the tree trunks which are difficult to be uprooted. These unwanted trees become a major problem in the orchards, as their roots and shoots get intermingled with the fruit tree both from roots and shoots. Some of the root portions usually are left in soil after their removal with spade. These roots re-sprout very fast giving out more number of plants. Weeds thus can be defined as undesirable and unwanted plants/trees in the orchards.

Disadvantages:

- Weeds reduce the economic yield of marketable fruits by competing with fruit plants for water, nutrition, light, space and air, which are meant for the fruit plants.
- Weeds create difficulty while carrying out orchard operation like hoeing, irrigation and harvesting of fruits.
- The weeds are very competitive in nature and are better adapted to the areas, where they grow, example, baru grass (Sorghum halepense), dib grass (Typha latifolia) ar\d motha (Cyprus rotundus) arid are capable of suppressing the growth of fruit plants in early years of planting.
- Some climbers as weeds like chhibber (Cucumis trigonus) spread on the canopy of fruit plants and check the light penetration completely in the fruit plants. Most of the newly planted fruit plants get killed due to the over powering of Karari (Convolvulus aroensis) and chhibber (Cucumis trigonus) climbers.
- Some weeds due to their rhizomes in soil like baru grass (Sorghum halepense) and knots of motha (Cyprus rotundus) resist their eradication. Weeds like Itsit (Trianthema portulacastrum) and chulai (Amaranthus viridis) have very high rate of seed production. The term one year seeding and seven years weeding holds true in such weeds.
- Weeds play as alternate hosts to many serious pests/diseases. The harbour insects/pests

e.g., Cynodon dactylon (Khabal grass) and Cyprus rotundus (Motha) are alternate hosts of grass hopper, likewise the 'wild senji' and 'maina' are hosts of aphids. Parthenium hysterophorus (congress grass) act as alternate host for mealy bugs, which attack mango, guava, pear and many other fruit trees.

- Some weeds are injurious to animal and human health. Parthenium and Dhatura seeds are poisonous in nature.

Advantages:

Although weeds have many disadvantages and are usually eradicated at a cost from orchards. At the same time weeds do have some advantages too.

- Weeds are used as fodder for animals and some weeds are used as food for mankind. For example Chulai and Bathu are used to prepare 'Saag' and 'chhiber' fruits are used to make chatni and pickle.
- Weeds are used to check soil erosion: In orchards where the slope/gradient of soil is more, as in foot hills, weed like (Cynodon dactylon) (khabbal grass) checks soil erosion due to its mat type of growth its fine roots help in fast penetration of rain water by making the soil more porus. In arid irrigated areas, wind erosion is common and can be reduced by growing kahi (Saccharum spontaneum).
- Some weeds have medicinal value for example homeopathic medicine for parthenium allergy is prepared from parthenium itself.
- Datura is used for animal diseases. Dried plants of Fumaria parviflora (Pitpapra) are used for the purification of blood to cure skin disease. Whole plant extract of Phyllanthus niguri (Bhoomi aonla) is beneficial in curing Jaundice.
- Some weeds are used as breeding material for sugarcane.
- Improvement of soil fertility: Weeds particularly legumes like maina (Medicago denticulate) and wild senji (Melilotus parviflora) are helpful in improving the soil nitrogen by the addition of root nodules in the soil.
- Weeds are used in cottage industry. Leaves of Typha latifolia (Dib) are used for making furniture and baskets.
- Some weeds are very sensitive to some toxins and can be used as pollution indicators for example bathu (Chenopodium album) is very sensitive to H_2S_2 and SO_3. Water hyacinth (Eshornia) can be used to reclaim industrial effluents; it can absorb heavy metals such as mercury, lead, iron, copper and tin etc.

Noxious Weeds

Weeds that parasitize crops form an important branch of weed science referred to as parasitic or noxious weeds. There are different groups of parasitic weeds including holoparasites (totally parasite), obligate parasites (growth and development depend on the host), and facultative parasite (semiparasite). Research results have shown a considerable amount of progress in understanding their physiology and biochemistry behavior. Key weed species in this category are Striga spp., Rhamphicarpa fistulosa, Alectra spp., and Orobanche spp. A strong relationship exists between Striga spp. parasitism on maize, for instance, and low soil fertility status, in particular, nitrogen deficiency. Addition of nitrogen has been used as a solution to reduce the effect of Striga spp. on maize.

Weeds from the Orobanchaceae family are serious parasitic pests in crop production, e.g., Striga asiatica (L.) Kuntze, Striga hermonthica. Benth and S. aspera (Wild) Benth. There is another group of weeds that produce chemicals that are detrimental to crops such as P. hysterophorus with its allelopathic effect but Desmodium spp. has allelopathic effect on other weeds, e.g., S. asiatica and S. hermonthica. Allelopathic effect of Desmodium spp. to Striga spp. has been utilized to develop a push–pull technology, which is a climate smart weed management technology. Under the push and pull technologies the weeds are used to interfere with germination of parasitic weeds to enhance crop growth and they provide competition useful for increasing growth.

Invasive Weeds

An alien plant also referred to as exotic, introduced, foreign, non-indigenous or non-native, is one that has been introduced by humans intentionally or otherwise through human agency or accidentally from one region to another. An alien plant that has escaped from its original ecosystem and is reproducing on its own in the regional flora is considered a naturalized species. Those naturalized aliens that become so successful as to spread in the flora and displace native biota or threatens valued environmental, agricultural or personal resources by the damage it causes are considered invasive.

Humans have been transporting animals and plants from one part of the world to another for thousands of years, sometimes deliberately for social or personal gain and sometimes accidentally. In most cases, such introductions are unsuccessful, but when they do become established as an invasive alien species as "an alien species which becomes established in natural or semi-natural ecosystems or habitat, is an agent of change, and threatens native biological diversity" the consequences can be catastrophic. Invasive alien species are the second largest cause of biodiversity loss in the world and impose high costs to agriculture, forestry, and aquatic ecosystems. In fact, introduced species are a greater threat to native biodiversity than pollution, harvest, and disease combined.

The global extent and rapid increase in invasive species is homogenising the world's flora and fauna and is recognized as a primary cause of global biodiversity loss. Bio-invasion may be considered as a form of biological pollution and significant component on global change and one of the major causes of species extinction.

Foresters, taxonomists and ecologists are now well aware of the problems caused by the invasion of alien species into natural areas and the associated negative effects on global patterns of native biodiversity. Once established, some alien species have the ability to displace or replace native plant and animal species, disrupt nutrient and fire cycles, and cause changes in the pattern of plant succession. Studies are underway to better understand the impacts of these species on native ecosystems.

Many invasive plants continue to be admired by people who may not be aware of their weedy nature. Others are recognized as weeds but property owners fail to do their part in preventing their spread. Some species do not even become invasive until they are neglected for a long time. Invasive plants are not all equally invasive. Some only colonize small areas and do not do so aggressively. Others may spread and come to dominate large areas in just a few years. The loss due to invasive species in United States estimated to be $125-150 billion each year and 25% of US agriculture gross national product lost due to foreign pests and weeds. Alien plants can spread rapidly because of our mobile society and the intentional transportation of ornamental and forage plants.

Characteristic Features of Invasive Species

Invasive species possess characteristic features like "pioneer species" in varied landscapes, tolerant of a wide range of soil and weather conditions, generalist in distribution, produces copious amounts of seed that disperse easily, grows aggressive root systems, short generation time, high dispersal rates, long flowering and fruiting periods, broad native range, abundant in native range. Preliminary data from one interesting study shows that invasive species are likely to have relatively small amounts of DNA in their cell nuclei. Apparently, the cells in these plants are able to divide and multiply more quickly and consequently the entire plant can grow more rapidly than species with higher cellular DNA content. This gives them a leg up in disturbed sites.

Process of Plant Invasion

The process involves five phases. Entry and spread of a weed species in a new geographic range has to tide over a series of physical barriers. Weed species that fail in this first phase in general do not acquire the status of quarantine pests by virtue of their dismal score in entry potential. If a species is successful in overcoming these barriers and gain entry in to a new range, the species passes through the phase of Entry/Escape and gets into the second phase of Establishment. In this phase, the weed species has to get through several abiotic and biotic stresses to establish itself in the new range. The species should be able to withstand, tolerate and come out of these biotic and abiotic stresses with self-sustaining populations, to establish itself, in the introduced/entered loci. Thereafter, these weed populations start expanding from this introduced loci to other surrounding habitats and this part of the process is the phase of Explosion, where the species concerned invades quickly large areas and several habitats. The last phase is the phase of Entrenchment, where the few left over habitats in the introduced range is being occupied by the weed and pattern of spread of the weeds starts plateauing, as the habitat range is saturated.

Factors Favouring Invasion

Fortunately, most introduced plants do not become invasive. 'tens' rule of thumb of biological invasions suggests that only 10% of introduced species will become established in a host environment and that only 10% of the established invaders will become pests. The extent to which an introduced plant naturalizes and spreads depends on the suitability of the new physical, chemical and biological environment in which it finds itself. If these factors are unsuitable, the plant is unlikely to become established. If some aspects of the environment is unsuitable, the plant may persist until there is an environmental shift in its favour or perhaps until it evolves to meet the adverse condition. There are several examples of weeds undergoing such a lag phase after introduction, later becoming significant noxious weeds. A third requirement for establishment and spread of an introduced plant may be a suitable disturbance regime, which creates appropriate open niches for their invasion and naturalization.

Genetic Variability

Any weed introduced in to a new environment evolves in to a rapid invasive species, when the genetic variability is comparatively lesser. Rubus alceifolius with higher magnitude of genetic variability in its native range (North Vietnam to Java) evolved itself as an invasive species in Madagascar with lesser genetic variability. This is a major weed in other Indian Ocean Islands, where its genetic variability is the least. Genetic introgression from Rhododendron catawbiense has been shown to impart cold tolerance on R. ponticum, a frequent invader in Britain and Ireland.

Climatic and Edaphic Factors

Casuarina equisetifolia has spread in California after the incidence of two hurricanes. In Cook Islands, the invasion of Cardiosperunum grandiflorum is reported to have been triggered by hurricane sally in 1986. Flooding is an important factor contributing for the invasion of some species with hydrophytic adaptations like Marsilea quadrifolia in India. Disappearance of physical barriers such as rivers, roads and cliffs or a secondary human introduction favours invasion by asexually reproducing species such as Selaginella in East Africa.

Biotic Factors

It is observed that introduction of a pollinating agent Parapristina verticillata (a fig wasp) was responsible for the invasion of Ficus microcarpa in Florida, USA that was planted as an ornamental originally, 45 years before. This pollinator is also reported to be responsible for the spread of the species in New Zealand. A shift in land use pattern in Hawaii from sheep grazing to cattle ranching allowed the shrub Ulex europaeus L. to invade. Cattle trampling resulted in microsites for seedling establishment. Cattle didn't graze on the species as effectively as sheep.

Absence of natural enemies in the introduced range is another important factor contributing for the invasive potential of weeds as in the case of spread of Eichhornia crassipes in several parts of the world; invasion of Acacia longifolia in South Africa and Chrysanthemoides monilifera in Australia. Allelopathic potential of the invading species against the coinhabitants in the introduced range is another factor. Companion species of Centaurea diffusa in the native range have evolved resistance to this allelopathic aggression, stalling the predominance of C. diffusa. However, in the introduced range susceptibility of the companion species to this allelopathic potential allowed the weed C. diffusa to become invasive. Interaction between the biotic factors and climatic factors might also encourage weed invasion.

The fact that the process of invasion of a plant species is largely dependent on its interaction with external factors whether biotic or abiotic, makes the need for observing the role of such factors on any potential invader, imperative to be forewarned.

Impact of Climatic Parameters of a Locality on Weed's Invasive Behaviour

Atmospheric Temperature

Invasive species whose native ranges are warmer than their introduced ranges would be at an advantage as they withstand hot temperatures better than natives of that region and as they suffer less mortality due to extreme cold conditions.

Increasing atmospheric temperatures could trigger invasion by some weeds in warm season crops. A rise in temperature by 3 °C of average temperature was shown to enhance biomass and leaf area of itch grass by 88% and 68%, respectively. The weed has also been reported with projected increases in maximum growth for the middle Atlantic States. This invasive weed causing significant yield reductions in sugarcane at Lousiana is also shown to be competitive in corn, cotton, soyabean, grain sorghum and rice systems. As many of these invasive weeds of southern U.S had their origin in areas associated with warm temperature of tropical climate, increasing temperatures might accelerate expansion of these weeds to northern states or region. Increasing temperatures could also trigger northward migration of cogongrass (Imperata cylindrica) and witch weed (Striga asiatica). Increase in temperature due to global warming might trigger plant migration exceeding the rate of the same observed in post-glacial period. This would lead to preferential selection of plant species with higher mobility. Plant traits that aid their dispersal over long distances are frequent in invasive plants. Accordingly, these invasive plants would be greatly triggered for migration by increasing temperatures.

A hypothesis predicted migration of Kudzu (Pueraria lobata gaining entry to south eastern United States from Japan) to northern parts. The weed is shown to be restricted by low winter temperatures of -15 °C, two decades ago. Recent observations indicated midwestern populations of this weed and this migration is attributed to increase in minimum winter temperature.

The impact of temperature changes on floristic composition is through tilting of balance towards crop in the process of crop weed competition. Crop losses due to weeds left uncontrolled are considerably higher in the south than in the north of U.S and this is attributed to the prescence of perennial invasive weeds like itch grass that are restricted in the northern states by low winter temperatures. The crop-invasive weed competition could also be influenced by the varying magnitudes of response to the interaction of rising temperature and floral reproductive ability. In Australia, the increase in mean temperature is expected to be associated with range extension of tropical prickly acacia A. nilotica to south.

Higher mean annual temperatures are shown to favour assimilate partitioning towards root biomass in the introduced exotic species P. juliflora, in Southern India. Greater root biomass of this species aids rapid and robust regeneration after lopping of the shrub for the popular fuel wood purpose serving the livelihood of rural poor in the drier areas or after revival of ecological stress conditions such as drought or submergence. Researchers suggest climate change with particular reference to increasing temperature would increase the risk of population establishment of new weed species in northern regions of Europe. A definite warming trend in Germany is expected to increase the chances for the establishment of species with a higher temperature optimum such as Vallisneria species and those intolerant to freezing such as P. stratiotes, Eichhornia crassipes and Salvinia auriculata.

Soil Temperature

Soil temperature plays a key role in promoting seed germination. Soil temperature could mainly serve the purpose of thermal induction of seed germination. Carpetweed Trianthema portulacastrum invading irrigated upland ecosystems in several tropical Asian countries is shown to be more invasive with synchronised mass seed germination, with increasing soil temperatures, in India.

Impact of Rainfall and Evaporation on Invasivity of Weeds

Rainfall and evaporation pattern of a region influences the weeds directly by interrupting the physiological functions involved in the process of seed dormancy and germination and by imparting seed mortality due to excessive drying or soaking. These effects are transferred through soil moisture status. The indirect impacts could be through displacement of seeds/propagules by run off, scorching because of removal of soil cover and mobility of nutrients for better stand establishment.

Soil Moisture

Invasive weeds of rangelands like cheat grass (Bromus tectorum) and yellow star thistle (Centaurea solstitialis L.) depend largely on available soil moisture for seed germination. Prolonged or heavy winters that greatly enrich the soil moisture favour increased seed production of both the species. Interestingly, both species are also drought adapted. The adoption of cheatgrass for drought evadance is shorter life span whereas, deeper root system to endure drought helps star thistle, compared to other native species. Increase in snowfall or a change in snowfall variability is suggested to exacerbate the invasion of forbs in mixed-grass prairie ecosystem which in turn could influence the forage availability. Soil moisture stress promoted greater herbivore tolerance for the invasive Alternanthera philoxeroides and decreased it for the native congener A. sessilis in China.

Invasive Weeds and Climate Management

Organisms, including plant species that have been moved from their native habitat to a distant location are typically referred to as non-native, nonindigenous, exotic, or alien to the new environment. An invasive species is one that both spreads in space and has negative impacts on species already in the space that it enters.

In the United States, an "invasive species" is defined by the Executive Order in 1999 as a species that is i) non-native (or alien) to the ecosystem under consideration and ii) whose introduction causes or is likely to cause economic or environmental harm or harm to human health. The Order further provides that a Federal agency may make a determination that the benefits of an action, which may lead to the introduction or spread of an invasive species, clearly outweigh the potential harm caused by the species and take steps to minimize that harm.

Invasive Weeds and Climate Change

The climate change is real whether you agree with or not. On a global basis, the current level of CO_2 is rising and temperature is slowly rising as well. It also indicates that we can alter global ecosystems such as the atmosphere, as the carbon released from energy use and other human activities triggers climate change. Plants are necessary for the flow of energy and carbon through ecosystems. With the exception of a few subterranean organisms, if plants did not exist, life would not exist. Plant growth, however, is dependent on four abiotic inputs: sunlight, nutrients, water and carbon dioxide. Any change in these inputs will alter all living systems.

Some evidence exists that agronomic weeds may reduce crop yields further in a higher CO_2 environment. It is also evident that rising CO_2 may be a selection factor in weed species dominance and its management strategies. Although the response of crop plants to rising atmospheric carbon dioxide concentration CO_2 has been well characterized, little is known concerning the long-term growth and/or photosynthetic response of perennial weeds. Available data indicate that sustained stimulation of photosynthesis and growth in perennial weeds could occur as atmospheric CO_2 increases, with a reduction in chemical control effectiveness and potential increases in weed/crop competition. The basis for increased glyphosate tolerance at elevated CO_2 for these cohorts were unclear, but was not related to plant size at the time of glyphosate application. And we don't know a lot yet on the occurrence of glyphosate tolerance. Climate change with increased CO_2 and increased temperature may affect growth and development of many weed species. Evidence suggests that climate change increases risk of plant invasion.

Invasive Weeds and Biodiversity

Human induced biological invasions are occurring on a global scale and are beginning to blur the regional distinctiveness of the Earth's biota. That distinctiveness, which developed over the past 180 million years as a result of the isolation of the continents (termed evolution in isolation), has maintained biodiversity. When considered as a single phenomenon, biological pollution probably has had greater impacts on the world's biota than more widely known aspects of global environmental change such as rising CO_2 concentrations, climate change, and decreasing stratospheric ozone levels.

Plant invasions by non-native species have become a major environmental problem. Unlike chemical pollutants that tend to degrade over time and permit an ecosystem to recover, biological invasions tend to multiply and spread, causing over-worsening problems. Insidious effects of invasive non-native species include displacement or replacement of native plants and animals, disruptions in nutrient and fire cycles, and changes in the pattern of plant successions. In general, these foreign species can pose a risk to biodiversity when they naturalize and penetrate conservation areas.

On a global scale, invasions by non-native plants, animals, fungi, and microbes are believed to be responsible for greater losses of biological diversity than any other factor except habitat loss and direct exploitation of organisms by humans. Non-native species further threaten fully two-thirds of all endangered species. Non-natives are now considered by some experts to be the second most important threat to biodiversity, after habitat destruction. Native species have also been considered invasive when they spread into human made habitats such as farms or gardens.

Nature and Distribution of Invasive Species

The nature and distribution of invasive species has no geographical boundaries. All living organisms– bacteria, fungi, plants and other organisms-have evolved to live in specific areas on the Earth. Local climate, geology, soils, available water and other natural factors may influence plant or organism's invasion and subsequent establishment in a particular habitat. Models predict how invasive plant distributions will shift with climate change.

1950, the number of plant introductions into the United States was estimated to be at least 180,000. In 1975, it was estimated that at least 1,800 introduced plant species had escaped into the wild, with a large proportion establishing free-living populations. Currently, the Weed Science Society of America recognizes about 2,100 plant species as weeds in the United States and Canada, including about 65% of all weeds in the United States are non-natives, approximately 1,365 of the weeds recognized by the Weed Science Society of America are of foreign origin. This does not include most weeds of natural areas. Of the 6,741 plant species that are recognized as weeds somewhere in the world, only 2,063 species occur in the contiguous United States. Currently, it is estimated that there are 4,678 species of invasive plants in other countries that could still be introduced into the United States.

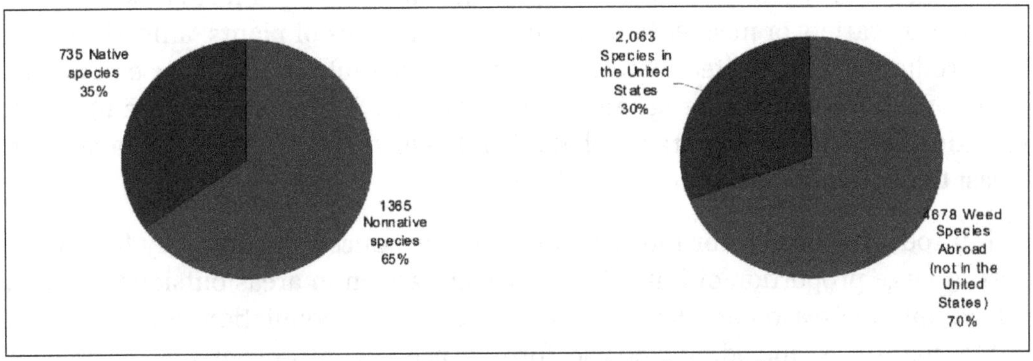

Numbers of native and non-native weeds in the United States.
Numbers of weed species in the United States and the World.

Attributes or Traits of Invasive Weeds

Invasive woody plants tend to have small seed size, a short juvenile period, and a relatively short interval between seed crops that produce a high number of seeds. Invasive plants (regardless of their origin) tend to have many similar biological attributes relating to high reproductive potential and stress tolerance. Some of the common n traits are as follows:

- Rapid seedling growth and early maturation.
- Ability to reproduce by vegetative propagules as well as by seeds.
- Ability to reproduce at an early stage ability to produce viable seeds and with seed dormancy ensuring periodic germination.
- Diverse dispersal mechanisms and high dispersal rate.
- Ability to tolerate wide range of environmental conditions.
- Ability to tolerate high habitat disturbance.

Invasiveness of plant species is dependent upon species traits, habitats, environmental stress and other biological factors. It has proven difficult to identify particular traits that are consistently associated with the tendency of plant species to invade. Knowing history of past invasiveness may be the best predictor on invasiveness of a species in a new habitat. This suggests that some plant species are more invasive than others but does not explain which traits encourage invasiveness.

Pathways of Introduction

Introduction means the movement, by human agency, of a species, subspecies, or lower taxa (including any part, propagule that might survive and subsequently reproduce) outside its natural range. Their movement can be either within a country or between countries. Plant invasions are important components of human-caused global environmental changes. However, researcher estimated that only 0.1% of all plant species that are introduced outside their native ranges by humans become invasive.

Invasive species have been introduced in a variety of ways. The means and routes by which they are introduced are called invasion "pathways". People introduce exotic plants to new areas with intent and by accident by variety of means. More than 50,000 species of plants, animals, and microbes have been introduced in the United States. Some species are introduced for use in gardening and landscaping, for erosion control, forage and other purposes. For instance, in the 1930's, the Civilian Conservation Corps planted Pueraria lobata (Wild.) Ohwi. introduced from Japan, throughout the Southeast to help stabilize soil in errodable areas.

Intentional introductions: Alien or foreign plants are introduced intentionally for a great variety of purposes. A large proportion of important crops are grown in areas outside their natural distribution for economic reason and as a way to feed the world population. Foreign plants can be introduced for many uses, including food use, forestry use, soil improvements, ornamental plants, cover crops and other uses. One classic example is Festuca arundinacea Schreb. a native European, has been planted as a pasture grass in North America. Now, it has naturalized and invaded remnant prairies, replacing the once diverse natural herbaceous community. Many introduced ornamental plants invaded in natural areas in the United States. The recent intentional introduction of Lythrum salicaria L. is another example of a successful plant invasion. This species was introduced to the Northeastern United States and Canada in the 1800s for ornamental and medicinal uses. Lythrum salicaria has adapted readily to both natural and disturbed wetlands. Various plant species also have been introduced as cover crops or for other purposes in India.

Accidental introductions: Many of the alien agricultural weed species have been accidentally introduced as contaminants of crop seeds. Despite the Federal Seed Act, weeds continue to arrive in the USA as seed contaminants. Similar contaminations of alien plant species have been identified in many countries in the world. Soil-inhabiting species can be introduced by shipping soil or by soil attached to plant material. Machinery and vehicles are often shipped from place to place without cleaning. Depending on their uses, they may carry soil and plant material. One of the best examples of accidental introduction is Phalaris minorin wheat fields of India. This accidental introduction reflects the importation of wheat from USA. Other invasive weed species such as Bromus tectorum, Parthenium hysterophorus and Lythrum salicaria will be highlighted at the conference.

Table: Intentional introduction of various weeds to India.

Scientific name	Common name	Introduced from	Purpose
Chromolaena odorata	Chromolina	Tropical areas	Cover crop
Eichnonia crassipes	Water hyacinth	South America	Ornamental
Lantana camera	Lantana	Many countries	Ornamental
Mikania micantha	Miconia	Malaysia	Cover crop
Opuntia strictsa	Opuntiaq	Australia	Hedge plant
Phaseolus labatus	Phaseolusq	USA	Cover crop
Sorghum halepanse	Johnson grass	USA	Forage crop

Economic Impacts

Invasive species are a major threat to the ecosystems. The differentiation of indigenous (native) and introduced (non-native) species often become difficult as only 10-15 percent of species on the earth has been fully identified. Though introduced species are often projected for their negative effects, introduced cultivated plant species contribute to 98% of human food supply equalling a monetary benefit of $5 trillion per year. However, the threats posed by non-indigenous invasive species often outweigh the benefits of their counterparts. The annual loss caused by invasive species (plant, animals and microbes) to global agriculture ranges from $55 to $248 billion. Invasive plant species can often turn to noxious weeds and cause considerable agricultural loss. The damage caused by weeds in global agriculture production comes to $94 billion, annually. In U.S., the agriculture weeds cause an economic damage of about $27.9 billion per year, while alien weeds inflict a damage of $37.8 billion annually in India. Lythrum salicaria after its introduction in North-eastern parts during the beginning of 20th. Century, has now acquired invasive proportion, and causes economic damage of $45 million annually in the United States.

According to the U.S. Congressional Office of Technology Assessment, there are at least 4,500 species of foreign plants and animals that have established free-living populations in the United States since the beginning of European colonization. Of that total, at least 675 species (15%) cause severe harm. In economic terms, 79 species, or 12% of total harmful species, caused documented losses of $97 billion from 1906 to 1991.

Potential Research on Invasive Species

We are beginning to understand invasive weeds/plants in relation to biology, ecology, invasiveness and biodiversity. However, there are many missing information/links as we attempt to predict environmental impacts of invasive plants. Integrated research needs for plant invasion with climate change are as follows:

- Growth and development of invasive plants n with climate change.
- Assessing reproductive potential of each invasive.
- Effects of rising co_2 on photosynthetic capacity.
- Effects of rising temperature on species adaptation.
- Use of biogeographic information on invasive nature and distribution.
- Use of gps on early detection program.
- Predicting plant invasion with climate change.
- Critical management needs for combating plant invasions with climate change.

Invasive plants have made significant changes to American landscapes over the last 200 years. Similar changes have taken place around the globe. India is no exception. Economic and environmental impacts of invasive species are enormous. There are many pressing issues regarding species invasion which must be addressed, including management strategies. Concerted efforts must be made to prevent the continued international movement of invasive plant species, and to develop programs to predict global changes. It is clear that invasive plants will continue to invade and persist in various ecosystems, and we must continue to monitor, assess, and develop management strategies for combating invasive species, and possibly to develop restoration of natural habitats. Again, we must continue to keep up with integrated current research on invasive plants with climate change, especially to rising CO_2 and temperature.

References

- Perennial-weeds, managing-weed-problems, weed-species, weedsmart: google.com, Retrieved 29, May 2020
- Weed-biology-part-1-monocot-vs-dicot: certifiedtraininginstitute.com, Retrieved 14, August 2020
- Woody-weed-control: cardinallawns.com, Retrieved 03, April 2020
- Grassland-weed-control: grassmanship.co.uk, Retrieved 25, January 2020
- Aquatic-weeds-meaning-and-control-management-fisheries-88528: yourarticlelibrary.com, Retrieved 17, June 2020
- Weed-management-in-orchards-types-methods-and-factors-affecting-efficacy-of-herbicides-24634: yourarticlelibrary.com, Retrieved 16, February 2020
- Noxious-weed, agricultural-and-biological-sciences: sciencedirect.com, Retrieved 23, July 2020

Chapter 3

Beneficial Weeds

The plants which are generally not cultivated and are wild but have some companion plant effect, contribute to soil health, are edible, or add ornamental value are the beneficial weeds. There are many plant species which are beneficial like Aegopodium podagraria, Allium vineale, Capsella bursa-pastoris, etc. This chapter discusses in detail these types of beneficial weeds.

Weeds are often the bane of the organic gardener's life. Even the most committed organic gardener can confess to being tempted to purchase a bottle of the chemical weed killer in order to get rid of the invasive creeping Charlie or nut's edge. While weeds are known to be annoying and to sap the nutrients from the soil, there are actually several beneficial weeds that every gardener should be aware of. Just like beneficial insects can have a positive impact on the health of your garden, so can beneficial weeds.

It is a well-known organic gardening fact that there are numerous plants and herbs that serve well as companion plants in order to keep pests at bay. These beneficial companions are typically planted on purpose in order to make a concerted effort at organic pest control. Seeds for these beneficial companions are found at most garden centers, and they are often considered to be "domesticated" varieties.

A beneficial weed, on the other hand, is a plant that is not typically thought of as being "domesticated," but it does still serve as a beneficial companion to your vegetables and flowering plants. Since a large number of gardeners are more concerned about pristine flower beds and lush green lawns that are free from the occasional pretty yellow dandelion, the majority of beneficial weeds are pulled up or poisoned with the harsh chemical weed killers.

The reality is that if these beneficial weeds were left in place, they would actually stave off hordes of unwanted pests and also help to keep the soil filled with valuable nutrients. With knowledge of what weeds might actually serve a positive purpose on your property, you can take your organic pest control and soil improvement efforts to extra lengths, all by allowing nature to work her magic.

Beneficial Weed Categories

Beneficial weeds typically fall into several categories, which can help you to define their best purpose on your own property.

- Repelling pests,
- Distractions or traps for pests,
- Soil fertilization,

- Ground cover,
- Natural herbicide,
- Attracting beneficial bugs,
- Useful for human use,
- Flavor enhancements.

Some beneficial weeds will actually serve multiple purposes, if you just know what to use them for. Once you know just how some of these beneficial weeds can improve your soil and your crop yields, you'll be a lot more likely to hesitate when you see a patch of dandelions growing in the middle of your tomato patch.

Pest Repellant

As organic gardeners, we know that planting onions and garlic near your lettuces, cabbages, carrots, and beets can help to keep pests away from your crops. The strong scents that onions and garlic plants put off will help to mask the delicious scent of your other crops, and thus protect them from being devoured.

Beneficial weeds work in much the same way—by masking the scent of the companion plants around them. Ground ivy, wormwood, wild oregano, and wild onions can work well to repel pests that would otherwise turn your tasty vegetables into their own dinner.

Some weeds are also useful to keep around because they have thorns or spines on them that can deter small animals. Rabbits and squirrels may seem impossible to deter at times, but you'd be surprised at just how effective a patch of thorn-covered weeds can be at keeping the critters away from your lettuce crop.

Pest Distractions

Insects track down their food by scent and color, which means that if you can effectively mask the scent of your tasty crops then you have won half of the battle. Some beneficial weeds work well as decoy plants that will not only attract the pests to them versus the tasty crops, but will also then often become the preferred meal over your crops.

Having a number of beneficial weeds like clover and ground ivy surrounding your crops has been shown to dramatically reduce numbers of tomato hornworms, squash bugs, and Japanese beetles. Patches of clover have also been known to distract the hordes of wild rabbits that would otherwise make short work of your cabbages and tasty young greens. Many an organic gardener has planted strips of clover as a type of organic line of defense. Keep in mind that some rabbits are just bound and determined to cross that line and make their way over to mow down the tops of your basil, but the clover should at least deter a fair number of them.

Soil Fertilization

We all know that planting beans amongst some of our other crops, like corn and squash, can increase the nitrogen content of the soil. Wild legumes can have the same effect on your crops, which means that white clover is actually a wonderfully beneficial weed to have growing in your garden.

Other weeds, like dandelions for example, have incredible root systems. They have extra-long tap roots that delve deep into the soil to bring up key nutrients that the other plants would otherwise not have access to. The quality of the soil will gradually improve over the course of a few years, but it would be constantly improving due to the presence of these soil-mining weeds. This also means that even if you have horribly dense clay soil, beneficial weeds like dandelions will ultimately transform the soil into something a lot more hospitable to vegetables and herbs.

Ground Cover

Think about the floor of a forest. This is one area that you would think would be densely populated with weeds and other unwanted plants. In reality, most forest floors are covered with thick ground-covering plants like ivy. Many beneficial weeds can be grown amongst your crops because they not only have different nutritional requirements from the soil, but they can also provide an effective ground cover that will essentially work as a type of living mulch. They will inhibit the growth of harmful weeds and will help to retain moisture in the soil. This can also often drop the temperature of the soil and help plants to better cope through the hottest and driest parts of the summer season. White and red clovers, along with rye grass, are often used to provide a nice ground cover for many crops.

Natural Herbicide

The plant world can actually be rather brutal, and the process of allelopathy is just one such example. Allelopathy is a process by which a plant produces biochemicals that will have a marked effect on the overall health of other plants around it. Allelopathy is not necessarily a positive process in all situations. For example, nut's edge can drive gardeners absolutely insane once it has been allowed to gain a foothold in the garden. Not only it is difficult to control, but it also releases toxins into the soil, thus making the soil inhospitable for other plants. This type of process can be beneficial for other plants, however. For example, many consider lantana a weed, but it can inhibit the growth of milkweed roots.

Attracting Beneficial Bugs

A garden that is teeming with beneficial insects is a healthy garden! There are dozens of beneficial weeds that can attract beneficial insects to them by offering nectar and a much more hospitable place for them to lay their eggs. An influx of more beneficial insects can only be a positive thing, as they take over pest control efforts on your crops. Attracting more ladybugs and predatory wasps is always a good thing for an organic gardener.

Useful for Human use

There are a number of uses for beneficial weeds, uses that go beyond the garden. Stinging nettles can certainly be irritating to your skin when you touch them, but you'll find that cattle and other livestock love eating them! As a matter of fact, you can eat them too! Nettles are incredibly high in a number of vitamins and minerals, including iron and calcium. They have a flavor that is very similar to spinach and can be cooked in much the same way. Nettles are often added to other herbs to create a vitamin-rich tea, so get a bit creative with your tea blends!

Dandelions are also definitely edible and are enjoyed in a number of dishes around the world. Dandelion leaves can be blanched to remove the bitterness, or they can be eaten fresh in salads and vegetable wraps. The petals of the flowers are used in the production of dandelion wine, while the roasted roots are often used to make dandelion coffee!

Purslane can also be enjoyed as a leaf vegetable, either in salads or stir-fried. All parts of the plant are edible, and purslane is often cooked much in the same way that spinach is cooked. It can also be added to soups and stews for the incredible nutritional content. Purslane is incredibly rich in omega-3 fatty acids and also contains high levels of iron, calcium, and vitamins A, C, and B.

Flavor Enhancement

Not only are some beneficial weeds edible in their own right, but studies have shown that a nice crop of stinging nettles planted amongst your mints and other herbs can actually increase the essential oil content of the herbs.

Dandelions have also been shown to improve the flavors of lettuces, tomatoes, and so many other vegetables around them. This can often be attributed to the dandelion's ability to improve the nutrient content of the soil, however.

Before you get ready to start pulling weeds out by the fistful, you should first do a quick check as to what you are actually growing. Nature may have planted some really beneficial weeds exactly right where you need them to grow.

Stinging Nettle

Stinging nettle, (Urtica dioica), also called common nettle, is a weedy perennial plant of the nettle family (Urticaceae), known for its stinging leaves. Stinging nettle is distributed nearly worldwide but is especially common in Europe, North America, North Africa, and parts of Asia. The plant is common in herbal medicine, and young leaves can be cooked and eaten as a nutritious potherb. Additionally, stinging nettle has been used as a source of bast fibres for textiles and is sometimes used in cosmetics.

Stinging nettle is an herbaceous plant and often grows to about 2 metres (6.5 feet) in height. The plant can spread vegetatively with its yellow creeping rhizomes and often forms dense colonies. The toothed leaves are borne oppositely along the stem, and both the stems and leaves are covered with numerous stinging and non-stinging trichomes (plant hairs). The plants can be dioecious (an individual produces only female or male flowers) or monoecious (an individual bears both male and female flowers), depending on the subspecies. The tiny green or white flowers are borne in dense whorled clusters in the leaf axils and stem tips and are wind-pollinated. The fruits are small achenes, and the plants produce copious amounts of seeds.

The stinging trichomes of the leaves and stems have bulbous tips that break off when brushed against, revealing needlelike tubes that pierce the skin. They inject a mix of acetylcholine, formic acid, histamine, and serotonin, causing an itchy, burning rash in humans and other animals that may last up to 12 hours. Hunting dogs running through stinging nettle thickets have been poisoned, sometimes lethally, by the massive accumulation of stings. This defense mechanism is an effective deterrent against most large herbivores, though the plant is important food for several butterfly species and aphids. The dried plant can be used as livestock feed, and heating or cooking the fresh leaves renders them safe for consumption.

Stinging nettle has a long history of use as a medicinal herb and is still used in folk medicine for a wide array of disorders, though there is limited clinical evidence supporting its efficacy. The rootstock is used as a diuretic and as an herbal treatment for benign prostatic hyperplasia (prostate enlargement) and other urinary disorders. Tea made from the leaves has been used to treat hay fever, diabetes, gout, and arthritis, and fresh stinging leaves are sometimes applied to arthritic joints in a process known as urtification, which is said to stimulate blood flow. Topical creams have also been developed for joint pain and various skin ailments, including eczema and dandruff.

Allium Vineale

This herbaceous perennial plant is 1-3' tall and un-branched. The central stem is round and hairless. Sometimes the lower stem becomes draped with the membranous sheaths of the outer layers of its skin and former leaves. These sheaths are brittle, light brown, and dried out. Along the lower half of the stem, there are a few alternate leaves that are up to 10" long and ¾" across at the base. They are linear and sword-shaped, hollow on the inside toward the base, and hairless. Relative to the stem, these fragile and rather floppy leaves are semi-erect, but they often bend at the middle or at the base. Each leaf is more or less round in circumference, becoming slightly flattened on the upper side toward the base, where it wraps around the stem.

The central stem terminates in a long naked stalk of flowers and/or aerial bulblets. Usually, a sessile cluster of aerial bulblets is produced; a few flowers on slender pedicels may appear above the bulbets. On rare occasions, an umbel of flowers will develop without bulblets. Regardless of whether flowers or bulblets are produced, the mature inflorescence is about about 2-3" across. During an early stage of development, the entire inflorescence is covered by a sack-like membrane that is rounded at the bottom and pointed at the top. This membrane splits open to release the bulbets and/or flowers. A dried remnant of this membrane usually persists at the base of the inflorescence. The bulblets are green to dark red, about 1/3" (8 mm.) long, and ovoid. Each bulblet has a long green tail that is about 1" long. The small flowers are green, white, or light purple. Each flower is about ¼" long, consisting of 6 tepals that are erect or slightly spreading. The blooming period can occur from late spring to mid-summer and lasts about 2-3 weeks. While the flowers are not noticeably fragrant, the foliage has a strong garlic aroma. Each flower is replaced by a 3-celled capsule containing several small black seeds. The root system consists of a bulb with secondary roots. Bulblets are often formed underground as offsets of the mother plant. This plant usually reproducesDistribution Map vegetatively by means of the aerial and underground bulblets, but it can also reproduce by seed when flowers are produced. This plant often forms colonies of variable size.

- Cultivation: This plant thrives in full or partial sun, moist to slightly dry conditions, and a heavy soil containing some clay. However, other kinds of soil are tolerated as well. This plant can spread aggressively and is difficult to get rid off. Occasionally, it invades lawns, especially when mowing is performed on an irregular basis.

- Range & Habitat: Field Garlic is a common plant that occurs in every county of Illinois. It is native to Eurasia. Habitats include prairie remnants and other areas along railroads, degraded meadows near rivers or woodlands, woodland borders and thickets, vacant lots, grassy clay banks, poorly maintained lawns, and waste areas. While this plant is usually found in degraded habitats, it also occurs occasionally in natural habitats, where it can become a pest.

- Faunal Associations: The nectar of the flowers attracts small bees and flower flies. Insects feeding on the bulbs of Allium spp. (Onions) include the maggots of Delia platura (Bean Seed Fly) and Delia antiqua (Onion Fly), while various Frankliniella spp. (Thrips) feed on the foliage. Although mammalian herbivores usually avoid this plant, it is sometimes eaten along with the grass by cattle. In the case of dairy cattle, this can result in milk with a disagreeable odor. The ecological value of this plant to fauna is low.

- Photographic Location: A grassy bank at the edge of the Red Bison Prairie in Savoy, Illinois. The inflorescences of the photographed plants are dominated by reddish aerial bulblets and their developing green shoots.

Field Garlic is the most aggressive Allium sp. in Illinois. Unlike the native Allium spp., it has leaves that are hollow at the base. Among the introduced Allium spp., some species produce only flowers, while others produce aerial bulblets primarily. Field Garlic belongs to this latter group. In this regard, it is similar to Allium sativum (Cultivated Garlic), but the aerial bulblets of Field Garlic have much longer green tails with a striking appearance. While its bulblets can be used to season food like Cultivated Garlic, they have a strong flavor.

Cruciferous Weed

Identifying weeds and understanding their growth habit can be a difficult, yet sometimes necessary task. Generally, to a gardener who prefers a tidy garden, a weed is a weed and needs to go, plain and simple. However, by identifying weeds, we can better understand how to control them. Not all weed control products or herbicides work the same way on every weed. The more you know about a specific weed, the easier it will be to choose the correct method of control.

These days, in the horticulture world, the term "cruciferous" is usually used to describe vegetables, such as:

- Broccoli,
- Cabbage,
- Cauliflower,
- Brussels sprouts,
- Bok choy,
- Garden cress.

These vegetables are considered cruciferous because they are all members of the Brassicaceae family. When discussing healthy eating, nutrition or super foods, leafy green cruciferous vegetables are very popular. In fact, cruciferous vegetables are the dominant crop all over the world.

Up until the early 20th century, plants which we now consider members of the Brassicaceae family were classified in the Cruciferae family. Both the present Brassicaceae family and past Cruciferae family include cruciferous vegetables; however, they also include hundreds of other plant species. Some of these other plant species are commonly known as cruciferous weeds.

How to Recognize Cruciferous Weeds

The words "Cruciferae" and "cruciferous" originate from crucifix or cross-bearing. The plant species that were originally classified in the Cruciferae family were grouped there because they all produced four petaled, cross-like blooms. Cruciferous weeds bear these crucifix-like blooms. However, these cruciferous weeds are actually members of the Brassicaceae plant family.

Weeds in the mustard family are sometimes called cruciferous weeds. Some common cruciferous weeds include:

- Wild mustard,
- Wild radish,
- Wild turnip,
- Hoary cress,
- Hairy bittercress,
- Pepperweed,
- Wintercress,
- Hesperis,
- Water cress,
- Bladderpod.

Many of the cruciferous plants that are considered invasive, noxious weeds in the United States originally came from Europe, Asia, North Africa or The Middle East. Most were considered a valuable food or medicine in their native regions, so early settlers and immigrants to the United States brought their seeds with them, where they soon got out of hand.

Cruciferous Weed Control

There are a number of methods that can be employed to help manage cruciferous weeds from the Brassicaceae family. Since their seeds can germinate throughout the year with sufficient soil moisture, keeping the area somewhat on the dry side may help. Pre-emergent herbicides, like corn gluten meal, can be used early on to help prevent germination.

For seedlings that do emerge, a post-emergent herbicide should be applied before the weeds become large enough to set seed. Burning, or flame weeding, is yet another option in suitable areas and with proper precautions taken. In areas where cruciferous weeds occur in low numbers, hand-pulling or spot spraying individual plants with an organic herbicide, such as vinegar or boiling water, may be a more preferable alternative.

Mimosa Pudica

Mimosa pudica is a weed which is commonly known as chuemue (touch me not) belongs to the family Leguminaceae. It is found in moist and waste places. Chuemue shows nyctynastic and seismonastic movement. All parts of this plant possess different medicinal properties like dysentery, leprosy, asthma, sedative, emetic and tonic. Its aerial parts contain different secondary plant metabolites like tannins, alkaloids, glycosides, Flavonoids and saponins.

Mimosa pudicaLinn (chuemue) is a sensitive plant belonging to the family Leguminaceae, distributed in Brazil and different parts of India. Mimosa pudica is well known for its rapid plant movement. The plant shows a peculiar nyctynastic movement i.e. the leaflets fold togetherin the evening and the whole leaves droops downward and then reopens at sunshine. The leaves also close up under various other stimuli such as touching, warming or shaking. The shrubs can also be transmitted to neighbouring leaves. This type of movement is called seismonastic movement. Mimosa pudica L. belongs to the genus mimosa (family: Fabaceae). Traditionally it is known to possess sedative, emetic, and tonic properties. All parts of the tree are considered to possess medicinal properties and used in the treatment of biliousness, leprosy, dysentery, inflammations, burning sensation, fatigue, asthma, leucoderma, blood diseases, alopecia, diarrhoea, insomnia, vaginal and uterine infections. It has spinous stipules and globose pinkish flower heads and grows as weed in almost all parts of the country. Sensitive plant has become a serious weed in fields of corn, soybeans, tomatoes, upland rice, cotton, bananas, sugarcane, coffee, oil palms, papayas, coconuts, and rubber in many tropical areas. It is particularly troublesome where hand pulling of weeds is practiced. The species may be controlled by a number of commercial broad-leaf herbicides.

Common names: Sensitive plant, humble plant, shameful plant, sleeping grass, touch-me-not. Other names: In Tongait is known as mateloi (false death);In Urdu it is known as Chui-Mue; In Bengali, this is known as Lojjaboti, the shy virgin; In Indonesia, it is known as PutriMalu (Shy Princess); In Malayalam it is called "Thottavaadi" (wilts by touch); In Marathi it is called "LazaLu" (shy); In Tamil, it is called Thotta-siningi (acts when touched); In Kannada, it is known as "Muttidare Muni" (angered by touch); In Malaysianit is called PokokSemalu (shy plant); In Burmese (Myanmar) it is called HtiKaYoan which means "crumbles when touched.

Geographical Location: Sensitive plant was first described from Brazil. It is a Common weed widely distributed in the Philippines in open, moist, waste places, open grasslands and open thickets, at low and medium altitudes in settled areas.

Morphological Characters

- Stem: It is reddish-brown in color, very much stiff, slender and growing to a length of 1.5 m (5 ft). It bears scattered thorns. The stem is erect in young plants, but becomes creeping or trailing with age.
- Leaves: The hairy leaves are alternate, bipinnately compound, with one or two pinnae pairs and 10-26 leaflets per pinna.
- Flowers: On close examination, it is seen that the floret petals are red in their upper part and the filaments are pink to lavender. Flowers have globose heads, peduncles prickly,

usually in auxiliary pairs all along the branches. Flowers are pollinated by the wind and insects.

- Fruit: The fruit is of straw colored consists of clusters of 2-8 pods of 1-2cm long each, prickly on the margins. The pods break into 2-5 segments and contain pale brown seeds 2.5mm long. The seeds have hard seed coats which restrict germination.

Phytochemical Investigation

The preliminary Phytochemical screening of chuemue showed the presence of alkaloids, flavonoids, saponins, tannins and phenolics, steroids, carbohydrates, resins, triterpenes, glycosides mainly c-glycoside. It also contains mimosine, mimosinamine, mimosinic acid, tyrosine 3, 4-dihydroxypiridine. The seed of the plant contains mucilage composed of d -xylose and d -glucuronic acid. They yield 17% of greenish yellow fatty oil. The plant contains tubulin which shows the ability to bind colchicine with its sulfhydryl groups. A new class of phytohormones-turgorines is active in the plant. These periodic leaf movement factors are derivatives of 4-O-(ß-D- glucopyranosyl- 6-sulphate) gallic acid. Different Flavonoids were isolated and identified as 5,7,3',4'-teteahydroxy-6-C-[β-D-apiose-(1→4)]-β-D-glucopyranosylflavones,5,7,4'-trihydroxyl-8-C-β-D-glucopyranosylflavones,isovitexin-2″-O-α-l-rhamnopyranoside,vitexin-2″-O-α-l-rhamnopyranosideand a mixture of flavonols avicularin and reynoutrin.

M. pudica is common in waste land and is also a weed of lawns, crops, pastures and roadsides. In Southeast Asia and the Pacific it is a serious weed in maize, sorghum, sugarcane, tea, soybeans, upland rice and cotton. Because of its tolerance to shading it is an important weed in plantation crops, such as rubber, coconuts, bananas, papaya, coffee, oil palm and citrus. In tropical pastures its dense growth and thorns often deter animals from feeding on suitable forage mingled with it.

Ethnopharmacology

The roots and leaves are commonly used in treatment as bitter, astringent, acrid, cooling vulnerary, alexipharmic, diuretic, antispasmodic, emetic, constipating and febrifuge. The leaves and roots are used in the treatment of piles and fistula. Paste of leaves is applied to hydrocele. Cotton impregnated with juice of leaves is used for dressing sinus. Plant is also used in the treatment of sore gum and is used as a blood purifier. In Ayurvedic and Unani system of medicine, this plant has been used in diseases arising from corrupted blood and bile, heart disorders, billious fever, piles, jaundice, leprosy, ulcers and small pox. The seeds and other plant parts of sensitive plant contain mimonsine, an amino acid that is known to cause hair loss and depressed growth in mammals.

Pharmacology

Ethanolic extract of leaves shows hypoglycemic activity in Alloxan-induced diabetic rats. Aqueous extract of Mimosa pudica shows antinociceptive activity. Methanolic extract showed significant ($p<0.05$) hepatoprotective effect by lowering the serum levels of various biochemical parameters such as serum glutamic oxaloacetate transaminase(SGOT), serum glutamic pyruvates transaminase (SGPT), alkaline phospatase (ALP), total bilirubin (TBL), total cholesterol (CHL) and by increasing the levels of total protein (TPTN) and albumin (ALB), in the selected model. Aqueous extract of dried roots of Mimosa pudica was tested for inhibitory activity on lethality, phospholipase activity, edema forming activity, fibrinolytic activity and hemorrhagic activity of Najanaja and Bangaruscaerulus venoms. Ethanolic extracts of Mimosa pudica leaves shows antimicrobial activity of the plant materials is due to the presence of active constituents like alkaloids or tannins. Mimosa pudica root methanolic extract showed very good wound healing activity when compared to the standard drug Gentamicin. The decoction ofMimosa pudica leaves shows anticonvulsant activity. The aqueous extract of Mimosa pudica is showed anti-asthmatic activity in in-vitro and in vivo animal models. Ethanol extract of Mimosa pudica exhibited a significant hypolipidemic activity. Alcoholic extract of Mimosa pudica significantly ($P < 0.001$) decreases the volume of gastric acid secretion, PH, free acidity, total acidity and ulcer index with respect to control. Ethanolic extract of roots of mimosa pudica Linn produced a significant and sustained increase in the aphrodisiac activity of normal male mice, without any diverse effects.

Goutweed

Goutweed (Aegopodium podagraria L.) is a common plant in the Apiaceae family. It occurs in Europe, North America and Asia. In folk medicine, mainly the leaves of goutweed were used to treat gout, inflammatory states in kidneys and bladder and to facilitate wound healing. Chemical composition of goutweed has not been entirely determined, however, it is known that the main group of compounds are polyacetylenes (falcarinol and falcarindiol). Moreover, other compounds were isolated from this plant, such as essential oils from the group of mono- and sesquiterpenes, as well as coumarins and polyphenol compounds, e.g. phenolic acids and flavonoids.

Aegopodium podagraria L. is a plant in the Apiaceae family. It commonly occurs in Europe, North America and Asia. Common names for this species are e.g. bishop's goutweed, bishop's weed, ground elder, herb Gerard, gout wort, wild masterwort, and snow-in-the-mountain. This perennial plant prefers shady and humid places. It can be seen in deciduous and mixed woodlands or forests, shrublands, gardens, near houses and shorelines. It is a weed difficult to eradicate, because is spreads quickly from root fragments and rhizomes. Aegopodium podagraria is highly competitive against other species and thus spreads rapidly on a given territory, often forming dense mats.

Chemical composition of A. podagraria is comparatively little known. Despite the studies on the content of some bioactive and mineral compounds, the information on the composition of goutweed is still incomplete, as often not all parts of the plants were analysed.

It could be deduced that the most important fraction of chemical compounds present in this plant are polyacetylenes (polyins). These compounds include falcarinol, falcarindiol, falcarinon and falcarinolon with anti-inflammatory and antimicrobial properties. One of the characteristics of Apiaceae family is the presence of aliphatic C17 – polyacetylenes of the falcarinol type. The substrates for the formation of these compounds are unsaturated fatty acids. Falcarinol and its derivatives are formed from oleic acid via oxidation and dehydrogenation. Polyacetylenes are present in all morphological parts of the plant (flowers, rhizomes, leaves, stems and seeds). Their biosynthesis increases in response to fungal, bacterial and viral infections. The contents of these compounds in certain parts of the plant changes depending on the time of the year. The highest amount is typically noted in leaves, however during blooming the flowers are the richest source of polyacetylenes. The content of falcarindiol is then several times higher than in other parts of the plant. It is assumed that it is related to protective function played by polyacetylenes. It was observed that essential oils from goutweed leaves contained 0.6% of falcarinol, whereas its stem contained 0.2%.

Table: Taxonomy of Goutweed.

Taxonomy	
Kingdom	Plantae – Plantes, Planta, Vegetal, plants
Subkingdom	Viridiplantae
Infrakingdom	Streptophyta – Land plants
Superdivision	Embryophyta
Division	Tracheophyta – Vascularplants, tracheophytes
Subdivision	Spermatophytina – Spermatophytes, seedplants, phanérogames
Class	Magnoliopsida
Superorder	Asteranae
Order	Asteranae
Family	Apiaceae
Genus	Aegopodium L. – Goutweed
Species	Aegopodium podagraria L. – Bishop's goutweed

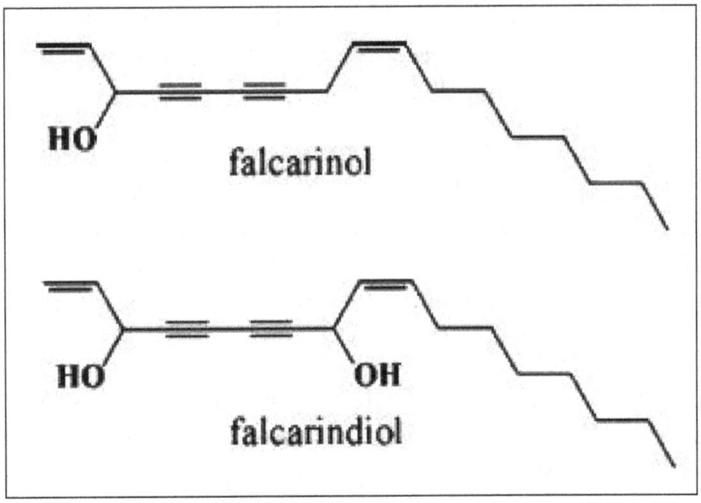

Chemical structure of main polyacetylenes present in Aegopodium podagraria.

Another important group of chemical compounds present in goutweed are essential oils containing mostly mono-and sesquiterpenes. Chemical composition of A. podagraria depends not only on vegetation period but also on the origin of the raw material. Goutweed from Russia (the Republic of Bashkortostan) during blooming was characterized by significant amount of sabinene at the level of 63%, and other essential oils were present at lower amounts, i.e.: α-pinene (3.6%), β-pinene (3.79%), myrcene (2.17%), α-thujone (0.63%), ethyl acetate (4.82%), dehydro-p-cymene (3.39%), and β-phellandrene present at the amount of 0.65%. Those compounds mainly belonged to the group of monoterpenes. In a plant from Siberia the prevailing substance in the essential oil was α-pinene (13.3%), and further on limonene (9.4%), p-cymene (8.8%), (Z)- β-ocymene (5.2%), β-pinene (5.0%), germacrene D (4.7%), spatulenol (4.4%), α-thujone (4.2%), perillaldehyde (4.1%),

(E)-β-caryophyllene (3.9%), (E)-caryophyllene oxide (3.4%) and myrcene (3.4%). The concentration of sabinene was only 1.8%. The differences between the composition of essential oils in leaves and stems of goutweed were observed in Estonia. More of these bioactive compounds were present in stems (3.8 mg/g) than in leaves (1.7 mg/g). In leaves, the main group was monoterpenes, including β-pinene (29.4%), limonene (18.4%), α-pinene (15.7%) and γ-terpinene (15.5%). In stems there were mainly monoterpenes (44%), dominated by β-pinene (11.1%), limonene (8.2%), γ-terpinene (8.2%) and α-pinene (6.6%), and in smaller amounts sesquiterpenes (29.8%), including germacrene (15.65%) and (E)-α-bergamotene (4.8%).

Goutweed also contains polyphenols from the phenolic acids, e.g. hydroxycinnamic acid, and flavonoids. These are the main compounds responsible for pharmacological properties of the plant. Jakubczyk et al. showed that that the overall content of polyphenols in the ethanolic extracts from different organs of Aegopodium podagraria L. depends on the temperature of extraction and the morphological part of the plant. The concentration range of polyphenols in the extracts ranged from 6.3 to 34.5 ppm. The highest polyphenol content was recorded for flower extracts, the lowest for seeds extracts. For all parts of the plant, with the exception of flowers, there was a higher content polyphenols in extracts prepared at 80 °C.

The Apiaceae family is also known to contain coumarins. Angelicin was found in leaves, fruits and roots of goutweed, and arterin was found in roots. The presence of angular furanocoumarins was also confirmed in leaves. In preparations from A. podagraria L. the derivatives of hydroxycinnamic acid were identified, e.g. caffeic acid and chlorogenic acid. Extracts from the plants contained flavonoid glycosides and derivatives of quercetin (hyperoside, isoquercetin), derivatives of kaempferol (trifoline) and unspecified rhamnoglucoside.

The analysis of carotenoids in the plant collected from April to May showed the changes in the content of analysed bioactive compounds. The amount of β-carotene ranged from 9.9 to 11.0 mg/100g of the raw material, depending on the time when the plant was collected, whereas the amount of xanthophyll was maintained on the level of 7.3–8.0 mg/100 g. The content of neoxanthin ranged from 4.1 to 4.7 mg/100 g. The content of lutein was higher in end May and amounted to 11.9 mg/100 g.

Goutweed is also appreciated as a source of vitamins and macro- and microelements, including iron, copper, manganese, calcium and potassium. Leaves, in comparison with stems, are a richer source of minerals (Zn, Cu, Cr, Mn, Co and Pb), excluding potassium, whose amount was twice lower than in stems (3.8 and 7.6 g/100 g of raw material, respectively). The plant contains 0.21-0.22 g/100g magnesium and also zinc (3.9 mg in leaves and 2.4 g in stems), manganese (3.2 mg in leaves and 2.0 mg in stems) and copper (0.37 mg in leaves and 0.19 mg in stems). In Dębia studies, the fluoride content in the infusions from this plant was measured. The concentration of fluoride in the infusions was diversified depending on the morphological part of the plant and the temperature of water used to prepare the infusions. The lowest content of this element was observed in rhizome infusions (0.057±0.004 mg/L), and the highest in seed infusions (0.333±0.045 mg/L). In both cases the infusions were prepared with water at 80 °C. The statistical analysis of the results showed that both the morphological part of plant and water temperature had an influence on the content of fluoride in the infusions. Ascorbic acid, tocopherols and carbohydrates (glucose, fructose) were already measured in A. podagraria L. In rhizomes, a high concentration of lectins, a group of proteins or glycoproteins binding carbohydrates, was also observed.

Properties and Folk use of the Plant

Aegopodium podagraria in folk medicine was used from ancient times, mainly in treatment of rheumatic diseases, gout and sciatica. The plant has calming, diuretic, anti-inflammatory and antimicrobial properties. Tea from dried leaves is a medicine recommended for the treatment of gout, haemorrhoids, inflammatory states of kidneys and bladder, auxiliary in nephrolithiasis and to improve metabolism. Fresh leaves administered on a wound facilitated quicker healing. Goutweed has not only therapeutic but also culinary significance. Young sprouts of goutweed were eaten in spring, cooked in soups, often with other plants. Most reports on goutweed consumption originate from the region currently belonging to Belorussia, Ukraine, Estonia, Scandinavian countries but also from Poland.

Anti-inflammatory Properties

The extracts at the concentration of 10mg/ml from rhizomes, stems, leaves and flowers were analysed for anti-inflammatory properties. The following solvents were used: water, methanol, hexane, acetone, dichloromethane and ethyl acetate. Falcarinol was extracted from plant material and its properties on the inhibition of inflammatory state enzyme, COX-1 (cyclooxygenase 1) were tested. It was proved that this substance had a 30-fold higher inhibitory effect on COX-1 than indomethacin, which is a synthetic drug used as an anti-inflammatory agent. IC_{50} value for the extract was only 0.3 μM, whereas for indomethacin 9 μM. The highest inhibition was noted for hexane extract from flowers. High level of falcarinol was observed in flowers during blooming, which could be related to protective effect of this compound during initial development of flowers. Later, the level of falcarinol decreased, therefore it is crucial to use anti-inflammatory potential of the flowers in initial period of blooming. It was shown that falcarinol inhibits the synthesis of COX-1, thus the extract of the plant may be used in the treatment of the diseases linked to inflammation. The results of Fujimoto et al. suggest that falcarinol (panaxynol) has the potential to inhibit PGDH activity in gastric mucosa, possibly as a result of pharmacological activity. At concentrations 25–200 μM falcarinol had no effect on the synthesis of prostaglandins (E2, F2α and D2) from exogenous arachidonic acid in the microsomal fraction of the gastric mucosa whereas at range from 25 to 200 μM it dose-dependently inhibited the activity of 15-hydroxyprostaglandin dehydrogenase (PGDH). IC50 was approximately 25 μM. Inhibition of PGDH by falcarinol was non-competitive with regard to NAD+ and prostaglandin E2. However, it is important to test the effect of goutweed extracts on the activity of other enzymes engaged in inflammatory processes, including cyclooxygenases 2 or lipoxygenases 5.

Antimicrobial Properties

Goutweed extracts exhibit antibacterial and antifungal properties. Available literature reports on weak antibacterial properties of goutweed extracts on Staphylococcus aureus, Staphylococcus epidermidis, Bacillus subtilis and Pseudomonas aeruginosa. It was also shown that extracts had antifungal activities. Delayed growth of pathogens was noted in case of Fusarium culmorum strains and Rhizoctonia sp., stronger growth inhibition in case of Heterobasidion annosum, and the strongest inhibition in case of Phytophthora sp. no inhibition was observed for Botrytis cinerea. In studies by Stefanovic et al. the strongest effect was exhibited by ethanol extract – MIC 5 mg/ml, whereas water extract did not have antibacterial properties. The antibacterial activity in vitro of water, ethanol and ethyl acetate extracts was also examined against 6 strains of bacteria. Antibacterial properties were determined by disc-diffusion and serial dilution methods. Based on minimal inhibitory concentrations (MIC) it was determined that ethanol extract showed the highest antimicrobial activity (1.25-5 mg/ml). Among the tested bacteria, the most susceptible were Enterobacter cloacae, Klebsiella pneumoniae and Pseudomonas fluorescens strains, whereas the most resistant was Staphylococcus aureus. Also, the interactions between ethanol extract and antibiotics – streptomycin and chloramphenicol, were examined. It was determined that the interactions between analysed parameters in people were synergistic, additive and neutral with respect to tested bacterial strains. Synergism was noted in case of Bacillus subtilis strains. Jakubczyk et al. showed that all tested extracts show inhibiting activity against reference and clinical strain of S. aureus with MIC value 64 mg/ml. Ethanolic extracts from Aegopodium podagraria L. because of the inhibiting activity can be helpful as addition to preparations used in the treatment of staphylococcal infections.

Studies suggest that coumarins, but also essential oils and flavonoids contained in goutweed, affect the plant's antibacterial and antifungal properties. Based on available literature it may be deduced that goutweed extracts varied in antimicrobial activity. This can be caused by extraction conditions, time and solvents used, let alone geological and climate factors, which affect the composition and properties of goutweed.

Antioxidative Properties

Aegopodium podagraria contains numerous biologically active substances with wide spectrum of activity. Antioxidative potential of the extracts from goutweed was assessed spectrophotometrically using ABTS and DPPH methods. The solvents used were chloroform, ethanol and ethyl acetate, and the final concentration of tested extracts was 1 mg/ml. The highest antioxidative potential (DPPH) was obtained for ethanol extract, where EC_{50} was 66.135 ± 1.6 μg ml^{-1}. For goutweed extracts this value was 25–125 μg ml^{-1}. The extracts prepared using chloroform and ethyl acetate showed weak antioxidant activity, less than 20% of inhibition of DPPH and ABTS radicals. It was thus confirmed that ethanol extract had antioxidative activity. The mechanism influencing those properties and the effect of certain parts of the plant remain unknown. Acquiring more knowledge in this area would enable to use goutweed as a safe diet supplement supporting antioxidative system of the organism and the use of its properties in treatment and prevention of many diseases linked to oxidative stress, including inflammation or atherosclerosis.

Effect on Metabolic Diseases

The metabolic effect of goutweed preparations was tested on the activity of metformin antihyperglycemic agent used in people suffering from type 2 diabetes and dexamethasone a type of corticosteroid medication, used in the treatment of many conditions, including rheumatic problems.

The metformin increases the glucose transporters (GLUT) transport abilities over a plasma membrane. It also exhibits a mild positive effect on lipid profile. Combination of the drug with plant material showing synergistic effect would enable to reduce the dose of the drug. In the study, the effect of tinctures on glycaemia and lipid profile was examined in rats who were given metformin and dexamethasone. In these studies used intact control and the test groups were given: metformin, dexamethasone + tap water, dexamethasone + metformin, dexamethasone + goutweed tincture (1 ml/kg), dexamethasone + metformin + goutweed tincture (1 ml/kg). Dexamethasone was used to induce a type 2 diabetes model. For 5 days dexamethasone was used in concentration of 5 mg/kg, whereas the concentration of metformin was 50 mg/kg. Insulin tolerance, glucose level and triacylglycerols, total cholesterol, HDL and LDL concentrations were analysed. It was confirmed that goutweed tincture increased sensitivity of rats to insulin and positively affected glucose level. In case of dyslipidaemia induced by dexamethasone, after administering the tincture the LDL fraction decreased, HDL fraction increased, and total cholesterol and triacylglycerols levels were reduced. Normoglycaemic effect may be linked to the presence of chlorogenic acid and ferulic acid in the tincture, which have a synergistic effect on metformin. Chlorogenic acid is a bioactive compound facilitating reduction of obesity. It increases the expression of GLUT4 via a pathway independent on PI3K kinase, whereas the activity of ferulic acid is PI3K – dependent. Thus, it can be stated that molecular mechanism in case of these relations is different and requires further studies.

In further studies the relation between the dose of goutweed extract and tincture was assessed on the metabolism of glucose in rats who were given metformin. The tincture showed hypoglycaemic effect in rats after administration of single doses of 0.5, 1.0 and 5.0 ml/kg, whereas the extract (100, 250, 500 mg/kg and 1.0g/kg) did not produce such effect. The results of correlation analysis indicated that there is some relation between the effect of the extract on electrolyte homeostasis and glucose metabolism, as well as between the activity of the preparations on uric acid and glucose level. The tincture at the doses of 0.5 and 1.0 ml/kg did not block the effects of metformin and did not increase its influence on excessive reduction of glucose level in blood. In glucose tolerance test, the tincture at the dose of 1.0 ml/ kg (but not 0.5 ml/kg) together with metformin reduced effective dose of the drug from 400 mg/kg to 200 mg/kg.

In contradiction to the extract, the tincture from goutweed shows hypoglycaemic effect, dependent on the dose, and seems to be a promising choice in connection with metformin. The nephroprotective and hepatoprotective effects of the goutweed extracts were also confirmed.

Preparations from goutweed seem to have the use in the treatment and prophylaxis of metabolic diseases, e.g. dyslipidaemia, type 2 diabetes. It is advisable to perform a more detailed analysis on the interactions of Aegopodium podagraria with metformin.

Cytotoxic Properties and Safety of Preparations from Aegopodium Podagraria

Aegopodium podagraria as a natural plant produce may be a valuable source of bioactive compounds facilitating the activity of our organism. Low level of toxicity of goutweed preparations was experimentally confirmed. Wide range between the therapeutic and toxic doses for the main active compound – falcarinol, allows to use the produce without the risk of side effects. Studies on falcarinol confirmed cytotoxic effect in case of some tumour lines, however, these were not the studies performed using goutweed extracts but only the substance isolated from other plant species. Despite the research on the toxicity of falcarinol present in other plant species of Apiaceae family there are no reports on the effects of goutweed extracts on particular cytotoxic parameters examined in vitro and in vivo.

Capsella Bursa-pastoris

Capsella bursa-pastoris is a annual growing to 0.3 m (1ft) by 0.1 m (0ft 4in). It is hardy to zone (UK) 6 and is not frost tender. It is in flower all year, and the seeds ripen all year. The species is hermaphrodite (has both male and female organs). The plant is self-fertile.

It is noted for attracting wildlife. Suitable for light (sandy), medium (loamy) and heavy (clay) soils. Suitable pH: acid, neutral and basic (alkaline) soils and can grow in very acid and saline soils. It can grow in semi-shade (light woodland) or no shade. It prefers dry or moist soil.

Common Name	Shepherd's Purse
Family	Brassicaceae or Cruciferae
USDA hardiness	6-10
Known Hazards	⚠ Signs of toxicity are sedation, pupil enlargement and breathing difficulty. Avoid if on treatments for high blood pressure. Avoid with thyroid gland disorders or heart disease. Possible addictive sedative effects with other depressants (e.g. Alcohol). Avoid during pregnancy.
Habitats	Arable land, gardens, waste places etc, it is a common weed of cultivated soil.
Range	A virtually cosmopolitan plant, found in most regions of the world including Britain.

Synonyms

Thlaspi bursa-pastoris, Bursa abscissa, Bursa druceana, Capsella concava.

Habitats

Cultivated Beds.

Edible Uses

Leaves - raw or cooked. The young leaves, used before the plant comes into flower, make a fine addition to salad. The leaves are a cress and cabbage substitute, becoming peppery with age. Leaves are usually available all year round, though they can also be dried for later use. The leaves contain about 2.9% protein, 0.2% fat, 3.4% carbohydrate, 1% ash. They are rich in iron, calcium and vitamin C. A zero moisture basis analysis is available. The young flowering shoots can be eaten raw or cooked. They are rather thin and fiddly but the taste is quite acceptable. They can be available at most times of the year. Seed - raw or cooked. It can be ground into a meal and used in soups etc. It is very fiddly to harvest and utilize, the seed is very small. The seed contains 35% of a fatty oil. This oil can be extracted and is edible. The seedpods can be used as a peppery seasoning for soups and stews. The fresh or dried root is a ginger substitute.

Composition

Figures in grams (g) or miligrams (mg) per 100g of food.

Leaves (Dry weight):

- 280 Calories per 100g,
- Water: 0%,
- Protein: 35.6g; Fat: 4.2g; Carbohydrate: 44.1g; Fibre: 10.2g; Ash: 16.1g,
- Minerals - Calcium: 1763mg; Phosphorus: 729mg; Iron: 40.7mg; Magnesium: 0mg; Sodium: 0mg; Potassium: 3939mg; Zinc: 0mg,
- Vitamins - A: 21949mg; Thiamine (B1): 2.12mg; Riboflavin (B2): 1.44mg; Niacin: 3.4mg; B6: 0mg; C: 305mg.

Medicinal Uses

Shepherd's purse is little used in herbalism, though it is a commonly used domestic remedy, being especially efficacious in the treatment of both internal and external bleeding, diarrhoea etc. A tea made from the whole plant is antiscorbutic, astringent, diuretic, emmenagogue, haemostatic, hypotensive, oxytocic, stimulant, vasoconstrictor, vasodilator and vulnerary. A tea made from the dried herb is considered to be a sovereign remedy against haemorrhages of all kinds - the stomach, the lungs, the uterus and more especially the kidneys. The plant can be used fresh or dried, for drying it is harvested in the summer. The dried herb quickly loses its effectiveness and should not be stored for more than a year. Clinical trials on the effectiveness of this plant as a wound herb have been inconclusive. It appears that either it varies considerably in its effectiveness from batch to batch, or perhaps a white fungus that is often found on the plant contains the medically active properties. The plant has been ranked 7th amongst 250 potential anti-fertility plants in China. It has proven uterine-contracting properties and is traditionally used during childbirth. The plant is a folk remedy for cancer - it contains fumaric acid which has markedly reduced growth and viability of Ehrlich tumour in mice. A homeopathic remedy is made from the fresh plant. It is used in the treatment of nose bleeds and urinary calculus. The German Commission E Monographs, a therapeutic guide to herbal medicine, approve Capsella bursa-pastoris Shepherd's Purse for nose bleeds, premenstrual syndrome, wounds & burns.

Caper Spurge

Caper Spurge is annual, biennial and sometimes perennial. It is folkloristically called "moleplant" for its capacity to repel rodents, and it is often confused with Euphorbia lathyris, which can be also found in literature as caper spurge.

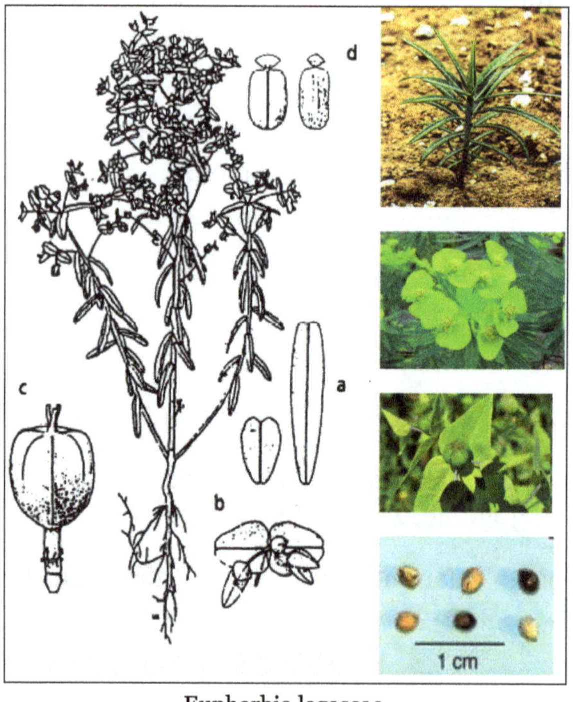

Euphorbia lagascae

The plant can reach 1 m tall and has smooth light green stems about 0.3-0.5 m long. Its structure is mainly simple and grows erect, but if it is disturbed and the apical dominance inhibited, the plant grows branching producing big amount of white sticky latex. It is succulent and leaves surfaces can be glaucous, covered with a greyish waxy coating or glabrous, smooth and glossy, without trichomes.

The leaves are ovate-linear below and narrowly lanceolate above, simple, entire, one-nerved, opposite, apically acute, basally abruptly truncate, long up to 15 cm long. The leaves subtending flower clusters broader. The flowers are subcordate, ovate-lanceolate and triangle lanceolate and greenish yellow. The fruits are usually 3-lobed, subglobose, up to 1 cm long. The capsule can harbour three or four seeds and it is ovate, acutely ribbed and scarlet in varied shades. The seeds are ovoid, flattened, and long up to 6 mm, black marked or speckled and hard to crush.

Domestication and Area of Origin

Native to the Mediterranean area, it is particularly present in Spain and Italy but less widespread than E. lathyris which can be found from Southern to Northern Europe, as well as in North America. E. lagascae has been introduced in North America and North Europe. The majority of researches were in Germany and the Netherlands. It occurs in the arid southeast and in the entire coastal region of Cadiz and grows spontaneously in cultivated field around towns. In Italy can be found as wild species in Sardinia.

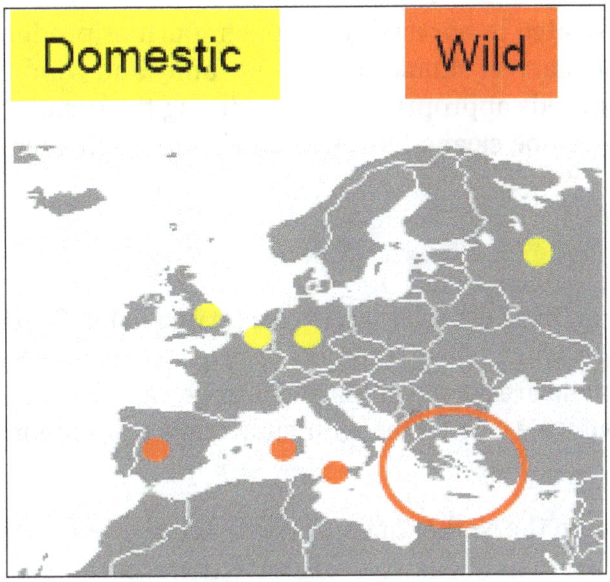

European distribution of E. lagascae.

Growing Conditions

The optimal sowing time is April to May. The seeding rate vary from 10 to 30 Kg/ha, 30 cm row spaced. Seed production seems to benefit from early sowing time and wider row spacing. The plant grows better in fallow lands rich in nitrogen, while it does not grow in siliceous or in poor soils. It seems tolerant to damp sites of saline nature.

Caper spurge can be considered a drought tolerant plant. Excess of irrigation water may generate too much greenness of stems, scalar seed maturation, massive release of latex and weed spreading. Otherwise, prolonged water stress periods may lead to vegetative regrowth in autumn when rainfall occurs more copiously.

Logistics: Harvesting/Handling

Caper spurge presents significant problems at harvest due to the scalar maturation. These problems comprise indeterminate growth and consequently a long period to reach capsule dehiscence. Moreover the foliage is rich in latex, that is vesicant and irritant and its handling can be critical, especially for the operators.

It is particularly important to define the right time of harvest. Common opinion is that harvest should be done when capsule ripening and seed dissemination has just started the third branch. Concerning the equipments various field researches on the harvest of Euphorbia lagascae have tested the suitability of different harvesting machines. Combine harvester seems not feasible due to high moisture content at harvest time, nor did the application of pre-harvest desiccants appear practicable. The best machine was the pea harvester, caper spurge being fairly similar to green peas. To avoid the blockage of the threshing drum, the machine can be equipped with a special cutter-bar table. Thanks to this tool, the sticky latex do not cause problems and no gum build-up occurs. During threshing, even if the process is rough and designed for pulses crop, there are no evident sign of seed damage.

Another critical process related to harvesting are the impurities produced by the capsule chambers. To reduce this issue, harvest should be done in sunny and windy days because dried capsules split and shed their seeds appropriately. After drying the organic matter up to a moisture content of 8%, the seeds can be cleaned through an air screen cleaner coupled with a brushing machine.

Production-Yields

Preliminary experiments on Euphorbia lagascae were carried out in Spain varying the date of sowing (autumn and spring), sowing density (36.000 to 143 000 plants ha–1), location (winter mild coastal and cold winter inland areas), irrigation and nitrogen application. Seed yield, was the highest in coastal locations with 0.4 to 1.6 t/ha with an average oil content of 44% (63% of vernolic acid).

In experiments carried out in the Netherlands, harvested seeds were 1.2 t/ha but the seed lost was remarkable, up to 0.4 t/ha, mainly due to dehiscence (Breemhaar and Bouman, 1995). Serious concerns with seed shattering were also reported in North Dakota where in spite of considerable seed yield potentials the harvested seeds amounted to only 0.1-0.2 t/ha.

Quality

The seed contains 46–52% of oil. The major fatty acid is vernolic acid (C18), which accounts from 60 to 75% of total oil content, mainly in triacylglycerols of seed cutin. It is the most known epoxy oleic acid from natural sources along with coronaric acid, that indeed is the principal fatty acid (up to 90%) of seed oil of Bernardia pulchella of the Euphorbiaceae family. Vernolic acid is an isomer of coronaric acid and both are biosynthetically related to linoleic acid. The epoxy acids are the best to produce paints and coatings, in fact vernolic acid is used as a drying solvent in alkyd resin paints, a plasticizer or additive in polyvinyl chloride (PVC) resins and possibly in pharmaceutical applications. Paints formulated with vernolic acid would greatly reduce volatile organic compound (VOC) air pollution that now occurs with volatilization of alkyd resins in conventional paints. On the other hand, the main menace of the poisonous seed oil of Euphorbiaceae is the presence of the same vernolic acid. Indeed, along with coronaric, these epoxy fatty acids are defined as "leukotoxins", generally referred to the acid produced by leukocytes in inflammatory status and for their primary toxic action against endothelial cells.

In the human body a little amount of vernolic acid is formed after epoxydation catalysis by cytochrome P450 enzyme and accumulates in the lung, in the vascular system and in neutrophils, interfering with the regulation of vascular tone, homeostasis and blood pressure. At high dosage the toxic cardiovascular effect could lead to death. This happen for example after severe burn, when the patient's skin is burn and leukocytes accumulate massively to limit skin infection, producing high amount of leukotoxin hypoxic hydrolases. This action is exerted by vernolic acid by-products, like its hydrated product, that resulted more toxic than the oxidized form. It is reported that vernolic acid by-products are the agent of the development of Acute Respiratory Distress Syndromes.

Applications: Current/Potential

Wild spurge has been grown for very different purposes, can be medicinal, ornamental or repellent for rodents in the orchards. The oil is the marketable products of wild spurge, nonetheless its industrial use is still little known. Along with medicinal uses, studies dates back to 1976 by Nobel Prize M. Calvin suggested to use it for biofuels, oil composition of wild spurge being very close to that of diesel oil.

The oil characterisation of E. lagascae revealed interesting potentialities for the production of soaps, detergents, lubricants, paints and cosmetics. The oil has high levels (up to 70%) of epoxy fatty acids, as vernolic acid that makes it suitable for epoxy coatings and resins.

Table: Applications of main compounds extracted from different organs.

Organ/compound	Applications
Seeds	Biofuels, adhesives, varnishes, paints, industrial coatings.
Oleic acid	
Vernolic acid	
Leaves	Pharmaceuticals, medical nutrition.
Quercetin	
p-coumaric acid	
Ferulic acid	
Latex	Pharmaceuticals, medical nutrition.
L-dopa	
β-sitosterol	

Additional uses are in the treatment of skin diseases, ulcers and warts, as well as cancer tumors and intestinal parasites. Some of the compounds of these plants are polycyclic and macrocyclic diterpenes with several biological effects such as, cytotoxic, antiviral multi drug resistant reversing activity, antitumor, antileishmanial and anti-inflammatory.

Restricting Factors

It is a highly toxic plant due to a poisonous milky latex. This substance primarily contain oligocyclic, polyfunctional diterpene esters present both in the seed and in the latex. It is highly skin irritant as well as possessing carcinogenic properties. This issue prevents the plant being used on a large scale, although breeding programmes are under way to reduce content of these substances. Due to the harmful nature of this specie for the presence of latex and other potentially irritating compounds in the stems and petioles, it will be important to understand which safety precautions are necessary during harvest and processing. Another important limitation for large scale production, that has hindered both breeding and agronomic research, is its massive seed shattering habit, making difficult to reduce seed losses during harvest and estimate and measure seed yield.

Moreover, the seed integrity is critical at harvest, because lipases activation within the seed change oil composition to the detriment of the epoxyoleic content. Last but not least, is the problem encountered with weeds. Indeed, weed control is a critical factor, because the common herbicides used are active even against cultivated wild spurge plants, resulting in vegetative damages and in lower seeds yield.

Edible Weeds

The following list includes some of the weeds that can be particularly useful, though not all plants listed should be tolerated due to their aggressive tendencies.

1. Aegopodium podagraria: Ground Elder is a vigorous and very invasive perennial, growing about 60cm tall and spreading rapidly by its roots. It is very difficult to eradicate because any small piece of root left in the ground will quickly regrow. Whilst it is occasionally grown as a ground cover in the wilder parts of the garden (shrubs and strong growing bulbs such as some lilies grow very well through it), it really is too vigorous for most other herbaceous species. There is, however, a variegated form of this species that is less invasive and is sometimes grown in the ornamental garden. Ground Elder has a long history of edible and medicinal use; indeed it was cultivated as a food crop and medicinal herb in the Middle Ages. It was used mainly as a food that could counteract gout, one of the effects of the rich foods eaten by monks, bishops etc at this time. The leaves can be eaten raw or cooked and have an unusual tangy flavour which is an acquired taste, although some people like it.

2. Allium vineale: Crow Garlic is a perennial bulb that grows up to 60cm tall. As a weed, it is most likely to be found in lawns that are not cut very regularly and spreads by means of the small bulbils that it produces instead of flowers. Both the bulbs and the bulbils have a rather strong garlic-like flavour -the bulbs can be harvested virtually all year round whilst the bulbils are ready in the summer. The leaves are also edible, with a similar flavour, but tend to have a stringy texture.

3. Capsella bursa-pastoris: Shepherd's Purse is an annual plant growing up to 50cm tall in rich soils, but much smaller in poor ones. A very common garden weed, it can flower and produce seeds all year round and will often spread freely in cultivated ground. This species is a prime example of how a plant can be viewed as an annoying weed in some areas of the world, whilst in others it is actually cultivated for its wide range of uses. It is extensively cultivated in some areas of the world as a cabbage-flavoured spring greens, whilst in Japan it is one of the essential ingredients of a ceremonial rice and barley gruel that is eaten on January 7th. The leaves grow rather larger under cultivation, they can be harvested about a month after sowing and can be treated as a cut and come again crop. The young leaves, used before the plant comes into flower, make a fine addition to salads, whilst older leaves are a cress and cabbage substitute, becoming peppery with age. The young flowering shoots can be eaten in the same way as broccoli. They are rather thin and fiddly but the taste is quite acceptable. The seed is rich in oil and, although small and very fiddly to harvest, can be eaten raw or cooked - it used to be ground into a meal and used in soups etc. The seedpods can be used as a peppery seasoning for soups and stews, whilst the fresh or dried root is used as a ginger substitute. The plant has long been used as a domestic herbal remedy, particularly in the treatment of both internal and external bleeding, diarrhoea etc. The seed, when placed in water, attracts mosquitoes. It has a gummy substance that binds the insects mouth to the seed. The seed also releases a substance toxic to the larvae - a kilo of seed is said to be able to kill 10 million larvae.

4. Cardamine species: Bittercress is a very common annual garden weed, growing just 1cm tall in very poor soils but up to 30cm in richer conditions. It spreads very freely by means of its seeds but is easily controlled by hoeing. It also tends to like growing in pots and it can be time consuming having to weed pot plants by hand. There are two main species found in gardens, the hairy bittercress C. hirsuta, and wavy bittercress C. flexuosa. Bittercress can be in flower all year round, especially if the winter is mild. Both the leaves and flowers can be eaten, raw and cooked. They have a hot cress-like flavour and can be used as a garnish or flavouring in salads, though they can also be used as a potherb.

5. Chenopodium album: Fat Hen is another annual weed that used to be cultivated as a food plant. Growing about 80cm tall, it was at one time often grown for its edible leaves. There is still one named variety that can sometimes be obtained - called 'Magenta' in reference to the colour of its leaves, it is considered by some people to be the best tasting of all potherbs. The leaves are a very acceptable spinach substitute, the taste is a little bland but this can be improved by adding a few stronger-flavoured leaves. The young flowering stems can also be eaten in much the same way as broccoli. The edible seed, which is somewhat fiddly to harvest, can be dried and ground into a meal and eaten raw or baked into bread. It can also be sprouted and added to salads. The seed should be soaked in water overnight and thoroughly rinsed before being used in order to remove any bitter-tasting saponins.

6. Cirsium species: There are many different species of thistle, the two you are most likely to encounter as weeds are C. arvense, the Creeping Thistle a perennial plant growing to about 90cm, and C. vulgare, the Common Thistle which is biennial and grows up to 2 metres tall. These are very aggressive weeds, the first spreading freely by means of its roots and the other sending its seeds far and wide to grow where you least expect them. The young roots of both species can be eaten raw or cooked. Although nutritious, they are rather bland with a taste reminiscent of Jerusalem artichokes. They are probably best when used in a mixture with other vegetables. Be warned though, just like Jerusalem artichokes the root is rich in inulin, a starch that cannot be digested by humans. This starch thus passes straight through the digestive system and, in some people, ferments to produce flatulence. The young stems can be peeled and cooked like asparagus or rhubarb. Young leaves have a fairly bland flavour and can also be eaten raw or cooked, but the prickles need to be removed before the leaves can be eaten - not only is this rather fiddly but very little edible leaf remains.

7. Elytrigia repens: Couch grass is one of the most pernicious weeds that gardeners have to contend with. It would be a very brave or very foolish person who would encourage this perennial grass into their garden. Growing about 50cm tall, the roots spread quickly through the soil, growing into the roots of other plants and making them almost impossible to get out. Even a small section of root left in the soil will quickly start growing again. Despite its antisocial tendency in the garden, however, couch is a very useful herbal medicine and the some herbalists used to say that half an acre of couch was worth five acres of carrots twice over. The edible uses of couch are very minor. The roots have been dried and ground into a powder, then used with wheat when making bread. Although thin and stringy, they contain starch and enzymes and are quite sweet. When boiled for a long time to break down the leathery membrane, syrup can be made from the roots and this is sometimes brewed into a beer. When roasted, the root has been ground into a powder and used as a healthier coffee substitute. Couch grass, however, is of considerable value as a herbal medicine -a decoction of the roots being very useful in the treatment of a wide range of kidney, liver and urinary disorders. They have a gentle remedial effect that is well tolerated by the body and has no side effects. This plant is also a favourite medicine of domestic cats and dogs, which will often eat quite large quantities of the leaves.

8. Epilobium angustifolium: With its tiny seed that floats so well on the air, rosebay willow herb will find its way to your garden without the need for you to invite it in. Left to its own devices, this perennial plant will then spread rapidly at the roots to form large clumps up to 2 metres tall. It produces spikes of flowers through much of the summer and, if it weren't so aggressive, it would surely merit a place in the ornamental garden.

The young leaves and shoot tips have been used in salads or cooked as a vegetable. When boiled they are said to make a wholesome vegetable and are a good source of vitamins A and C. The root can be eaten raw, cooked or dried and ground into a powder. When used in spring, it is said to have a sweet taste. The flower stalks, picked when the flowers are still in bud, can be eaten raw or cooked. The pith of young or older stems is slightly sweet, tender and quite pleasing to eat, though there is not much of it. A sweet and pleasant tea can be made from the dried leaves - it can be consumed on its own or mixed with conventional tea. Rosebay willow herb has often been used as a domestic herbal remedy, particularly to treat conditions such as diarrhoea and irritable bowel syndrome; it is used in Germany and Austria to treat prostate problems and a poultice of the leaves is applied to mouth ulcers. A poultice made from the peeled roots is applied to burns, skin sores, swellings, boils etc. A fibre obtained from the outer stems is used to make cordage, whilst the 'cottony' seed hairs are used as a stuffing material in toys etc or as tinder to start fires.

9. Oxalis corniculata: Yellow Sorrel is a low-growing annual or short-lived perennial plant, rarely more than 1cm tall but spreading to form clumps 15cm or more across. This is another weed we that can become a nuisance if it gets into pot plants, where it spreads rapidly by seed and underground

bulbils. Both the leaves and the flowers have a pleasant acid flavour, rather like sorrel. They are very small and fiddly to harvest, but can be added to salads. The leaves are used as an antidote to poisoning by the seeds of Datura spp, arsenic and mercury, and the leaf juice has a soothing effect when applied to insect bites, burns and skin eruptions.

10. Plantago major: Common Plantain is a common lawn weed. This perennial plant grows up to 20cm tall, though it will be much lower when growing in a frequently cut lawn. Although most gardeners mercilessly root it out of their lawns, it actually does no harm when growing there and, indeed, helps to maintain the fertility of the lawn. The young leaves are rather bitter and tedious to prepare because the fibrous strands need to be removed before use, but they have been used as a potherb. It is best not to use the leaf stalk since this is even more fibrous than the leaf. They can be blanched in boiling water before using them in salads in order to make them more tender. Although very tedious to harvest, the seed can be ground into a meal and mixed with flour when making bread, cakes etc. The whole seed can also be boiled and used like sago. Common plantain is a safe and effective medicinal herb. The leaves are used externally as a healing poultice and treatment for bleeding, quickly staunching blood flow and encouraging the repair of damaged tissue. Internally, they are used in the treatment of a wide range of complaints including diarrhoea, gastritis, peptic ulcers, irritable bowel syndrome, haemorrhage, haemorrhoids, cystitis, bronchitis, catarrh, sinusitis, asthma and hay fever. Plantain seed husks are an excellent treatment for digestive disorders. They contain up to 30% mucilage that swells up in the gut, acting as a bulk laxative and soothing irritated membranes.

11. Polygonum aviculare: Knotweed is an annual plant growing to 30cm in height. It is a common and invasive weed of cultivated ground, spreading by seed but quite easily controlled. It is an important food plant for the caterpillars of many species of butterflies and also produces an abundance of seeds that are a favourite food for many species of birds. Young leaves and plants can be used as a potherb; they are a very rich source of zinc. The seed, which is rather small and fiddly to utilize, can be used in all the ways that buckwheat (Fagopyrum esculentum) is used, either whole or dried and ground into a powder for use in pancakes, biscuits and piñole. Knotweed is a safe and effective astringent and diuretic herb that is used mainly in the treatment of complaints such as bleeding, dysentery and haemorrhoids. It is also taken in the treatment of pulmonary complaints because the silicic acid it contains strengthens connective tissue in the lungs. Recent research has shown that knotweed paste applied internally is a useful medicine in the treatment of bacterial dysentery. The plant yields a blue dye that is not much inferior to indigo, whilst yellow and green dyes can also be obtained.

12. Polygonum japonicum: Japanese Knotweed is not the sort of plant to encourage or even tolerate in the garden. A perennial growing 2 metres or more tall, it spreads very rapidly at the roots to form dense clumps that smother any other plant growing there. Since its introduction from the East, it has become one of our most pernicious weeds, so vigorous that the roots have even managed to send up shoots through tarmac! However, as with all plants, it is not all bad news. A report stated that Japanese knotweed is actually becoming a very valuable habitat for spiders, frogs, grass snakes and many other creatures. Its hollow stems allow a wide variety of insects and other small creatures to overwinter and find hiding places, thus a greater abundance of food is provided for insectivores such as frogs, who are themselves eaten by grass snakes. In areas of north Wales where Japanese knotweed has run rife, it is now the primary habitat for grass snakes. The plant also has several edible uses. The young shoots, which have a mild acid flavour, can be eaten in spring, either cooked in the same way as asparagus or used as a rhubarb substitute in pies, fruit soups, jams etc. The seed can be ground into a powder and used as a flavouring and thickener in soups etc, or can be mixed with cereals when making bread, cakes etc.

13. Prunella vulgaris: Self-Heal is a perennial plant, growing up to 15cm tall and often found as a weed in the lawn. The leaves have a somewhat bitter flavour but can be used in salads, soups, stews etc. The bitterness can be reduced or even removed by washing the leaves prior to use. A cold water infusion of the freshly chopped or dried and powdered leaves is used as a very tasty refreshing beverage. Medicinally, self-heal has a long history of folk use, especially in the treatment of wounds, ulcers,

sores etc. It was also taken internally as a tea in the treatment of conditions such as fevers, diarrhoea, sore mouth and internal bleeding. An olive-green dye can be obtained from the flowers and stems.

14. Pteridium aquilinum: Bracken is an extremely invasive plant and is a noxious weed. It is one of the most widespread plants in the world, being found in all parts of the globe other than the extreme north and south. A perennial plant, growing about 1.2 metres tall, it spreads freely both by the roots and its dust-like spores that can travel hundreds of miles in the air. The plant has a number of edible uses, with the roots and young shoots being eaten. However, there is also evidence to suggest that regular consumption of the plant could cause cancer, so eating it is not recommended.

There are also many non-edible uses. For example: a glue can be made from the rootstock; the rhizome lathers readily in water and can be used as a soap; the roots have been rubbed into the scalp in order to promote hair growth; the ashes of the plant are rich in potassium and can be used as a fertilizer and also in making glass; the dried ferns produce a very durable thatch and the fibrous remnants from edible roots make a good tinder. In addition, the leaves are used as a packing material for fruit, keeping it fresh and cool without imparting any colour or flavour. They can also be used as a lining for baskets, fruit drying racks etc where they help to repel insects and can also help to prevent rot in the fruits. The dried fronds are very useful in the garden as a mulch for somewhat tender plants. This will keep the soil warmer, protect from wind damage and also keep off some of the rain.

15. Rumex crispus: Curled Dock is a perennial plant growing about 60cm tall. It is considered to be a serious weed of agriculture but perhaps it should be tolerated as much as possible because it is such a good plant for re-establishing fertility in the soil -its deep roots bring up nutriments that would otherwise be lost whilst its leaves make excellent compost. It is also a very important food plant for the caterpillars of many species of butterfly.

Although rather bitter, the very nutritious leaves have been used in mixed salads, cooked as a pot-herb or added to soups. Only the very young leaves should be used, preferably before the stems have developed, and even these are likely to be bitter. The inner portion of the stems is also eaten whilst the seed can be used as a piñole or can be ground into a powder and used as flour for making pancakes etc. When roasted, the seed has been used as a coffee substitute. Curled dock has a long history of domestic herbal use. It is a gentle and safe laxative, less powerful than rhubarb in its action so it is particularly useful in the treatment of mild constipation. The plant has valuable cleansing properties and is useful for treating a wide range of skin problem. All parts of the plant can be used, though the root is most active medicinally.

16. Sonchus arvensis: Field Milk Thistle is a common garden weed that has occasionally been cultivated as a food plant -in Indonesia they have even produced improved varieties selected for their edible leaves. A perennial growing up to 1.2 metres tall, it spreads by means of seed and its creeping rootstock. The young leaves have a slightly bitter taste but can be added to salads or cooked like spinach. It might be best, though it is not necessary, to remove the marginal prickles. The mild flavoured stems can be cooked like asparagus or rhubarb, whilst the roasted root is used as a coffee substitute.

17. Stellaria media: A very common garden weed, chickweed grows, flowers and sets seed all year round. An annual plant growing about 15cm it spreads by means of seeds. It is very easy to control by hoeing and it has many beneficial uses. The young leaves have a mild flavour and can be available all year round if the winter is not too severe. Very nutritious, they can be eaten raw in mixed salads, or cooked to make a very acceptable spinach substitute. The small seed can be ground into a powder and used in making bread or to thicken soups. Chickweed has a very long history of herbal use, being particularly beneficial in the external treatment of any kind of itching skin condition. It has been known to soothe severe itchiness even where all other remedies have failed. When applied as a poultice, it will relieve any kind of roseola and is effective wherever there are fragile superficial veins. An infusion of the fresh or dried herb can be added to the bath water and its emollient property will help to reduce inflammation -in rheumatic joints for example - and encourage tissue repair.

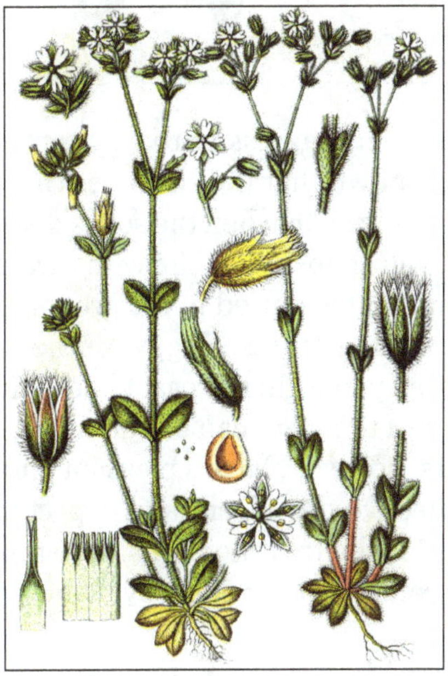

18. Taraxacum officinale: Dandelions are another of those weeds that is at times cultivated for its edible leaves - indeed there are a number of named varieties that have been developed in Europe. A perennial plant growing up to 40cm tall, it may be perhaps be encouraged in a lawn, but in cultivated beds it tends to provide a hiding place for slugs under the leaves. The plant spreads very freely by means of its light seeds, but is easily controlled by hoeing. The leaves are literally packed full of vitamins and minerals, making this one of the most nutritious leaves you can eat. Unfortunately, they have a rather bitter flavour, though we find a few of the leaves added to a mixed salad to be quite acceptable. The bitter tasting root can also be cooked; some people say they have a turnip-like flavour.

When roasted and ground into a powder, they can be used as a much healthier alternative to coffee. The dandelion is a commonly used herbal remedy. It is especially effective and valuable as a diuretic because it contains high levels of potassium salts and therefore can replace the potassium that is lost from the body when diuretics are used. The latex contained in the plant sap can be used to remove corns, warts and verrucae. The latex has a specific action on inflammations of the gall bladder and is also believed to remove stones in the liver.

19. Tussilago farfara: Coltsfoot has a very aggressive root system and will spread very freely if given half a chance. Indeed, it has been shown that the rhizomes can lay dormant in the soil for many years, finally emerging and resuming growth when the soil is disturbed. All in all, this is not a plant to encourage in the garden unless you choose your site with care or find some means of restraining it such as by planting in a large tub that is buried in the ground. A perennial plant, it grows up to 20cm tall, sending up flowering stems in the spring, with the leaves only appearing after it has finished flowering. This habit has led people in the past to believing that the flowers and the leaves were two different plants. All parts of the plant can be eaten, but most have a rather bitter flavour. The flower buds and young flowers, however, have a pleasant aniseed flavour and add a distinctive aromatic flavour to salads. An effective demulcent and expectorant herb, coltsfoot is one of the most popular European remedies for the treatment of a wide range of chest complaints, soothing coughs and helping to get rid of catarrh. It is widely available in health food shops, often in the form of a candied sweet.

20. Urtica dioica: If there is one plant that every child learns to recognise, it is the stinging nettle. Just one brush up against the stinging hairs on its leaves is enough to produce a lasting memory! A

perennial plant, it grows up to 1.2 metres tall and can form very extensive clumps spreading both at the roots and by seed. Apart from its many uses for humans, this is a superb wildlife plant. It can be a nuisance in cultivated areas, but it should be allowed to grow in out of the way places. Stinging nettles have a huge range of uses, only a few of which are mentioned here. The young leaves can be eaten cooked -although they can sting you when raw, once they are cooked they are perfectly harmless and make one of the most nutrient-rich foods you can eat. Do not harvest leaves from plants that have started to flower, however, since these can develop gritty particles called cystoliths that act as an irritant to the kidneys. If you cut the plants down twice a year, just before they start to flower, then you will ensure a good supply of fresh young leaves in the summer and will also prevent the clump from becoming too vigorous. Cutting them more than twice a year will severely weaken them and probably kill them over a period of 2-3 years.

Nettles have a very long history of herbal use and modern research has shown the justification for this. For example, a tea made from the leaves has traditionally been used as a cleansing tonic and blood purifier so the plant is often used in the treatment of hay fever, arthritis, anaemia etc. The roots contain substances that have been shown to have a beneficial effect upon the prostate and it is now used in a very effective formula to treat an enlarged prostate. The fresh leaves of nettles have been rubbed or beaten onto the skin in the treatment of rheumatism etc. This practice, called urtification, causes intense irritation to the skin as it is stung by the nettles. It is believed that this treatment works in two ways. Firstly, it acts as a counter-irritant, bringing more blood to the area to help remove the toxins that cause rheumatism. Secondly, the formic acid from the nettles is believed to have a beneficial effect upon the rheumatic joints.

A strong flax-like fibre obtained from the stems is used for making string, cloth and a good quality paper. The leaves are an excellent addition to the compost heap and can be soaked for 7-21 days in water to make a very nutritious liquid feed for plants. This liquid feed is both insect repellent and a good foliar feed. A hair wash is made from the infused leaves and this is used as a tonic and anti-dandruff treatment. A beautiful and permanent green dye is obtained from a decoction of the leaves and stems.

References

- The-basics-of-beneficial-weeds: offthegridnews.com, Retrieved 13, April 2020
- Stinging-nettle, plant: britannica.com, Retrieved 26, June 2020
- Field-garlic: illinoiswildflowers.info, Retrieved 18, March 2020
- Cruciferous-weed-information, plant-problems: gardeningknowhow.com, Retrieved 20, July 2020
- Review-on-mimosa-pudica-useful-weed: pharmatutor.org, Retrieved 10, February 2020
- Goutweed-Aegopodium-podagraria-L-botanical-characteristics-and-prohealthy-properties-339385815: research-gate.net, Retrieved 29, May 2020

Chapter 4

Weed Biology and Ecology

The study of the establishment, growth, reproduction and life cycles of weeds is known as weed biology. The various stages related to weeds like weed seed production, weed seed dispersal, weed seed germination and various aspects of weed ecology like weed seed longevity, weed crop competition and weed thresholds, etc. are discussed in an easy-to-understand manner in this chapter.

Weed Biology

Weed biology is the study of the establishment, growth, reproduction, and life cycles of weed species and weed societies/vegetation. Weed biology is an integrated science with the aim of minimizing the negative effects, as well as using and developing the positive effects, of weeds. Life Cycle - Based on life cycle weeds are classified as annuals, biennials or perennials.

Growth Characters

Based on growth characters weeds are classified as grasses, broadleaf weeds, ferns and herbaceous or woody weeds.

Broadleaf

Broadleaf weeds are annual, biennial or perennial plants which generally have two leaves (cotyledons) emerging upon germination. The leaves normally have a branching network of veins and the flowers have distinct petals.

Grasses, Sedges and Rushes

Grasses can be annual or perennial plants. They generally have narrow, upright, parallel-veined leaves. Grasses have jointed stems, usually hollow at the internodes and are circular in cross section. Sedges are a large group of perennial (rarely annual) grass-like plants which are common in wet, poorly drained soils. Sedge stems are triangular in cross section, solid, and not jointed.

Rushes are annual or perennial plants similar in appearance to sedges with grass-like tufted leaves common at the plant base. Rush stems are hollow, circular in cross section, and not jointed. Like the sedge, this plant is also common in wet areas or poorly drained soil, but is also found in woodland and open field.

Ferns

Ferns are primitive perennial plants that do not produce flowers and seeds. Ferns consist of a leaf or frond, a stalk and an expanded blade which may then be further subdivided several times. Ferns spread by long creeping rhizomes and/or by spores.

Fern Plant and Fern Spores.

Weed Reproduction

- Weeds most often reproduce by seed.
- Some weeds reproduce through vegetative means:
    - New individuals formed from multicellular structures of a single plant.
    - Vegetative reproduction is a trait that is shared by 60% of the world's worst weeds.
    - Reproduction by vegetative methods makes weed control extremely difficult.
- An understanding of a weed's breeding system, methods of propagule dispersal, and life form is important.
- These attributes can be used to predict population spread and longevity.
- They can also be used to predict genetic diversity within and among populations.
- The more genetically diverse a population:
    - The more it is buffered from management strategies.
    - More likely those individuals are present that can adapt to the strategy.

Breeding Systems in Plants

- Outcrossing,
- Self-pollination,
- Clonal,

- Mixed mating,
- Breeding systems for many weeds are not well known.

Outcrossing in Weeds

- The gametes that form the zygote are genetically dissimilar.
    - Self-incompatibility may be involved.
    - Outcrossing may occur between closely related individuals.
    - Inbreeding depression.
- Leads to more diversity within a population.
- Leads to less diversity among populations.
    - Populations from different geographies are not very specialized.
- At low densities there may not be enough pollen to produce the full potential of seed.
- If the species is an obligate outcrosser, at least two plants are required to start a new population.

Self-pollination in Weeds

- The gametes that form the zygote are genetically very similar.
- Leads to less diversity within a population.
    - A population can be relatively homozygous.
- Leads to more diversity among populations.
    - Populations from different geographies can be very specialized.
- If the species can self-pollinate, only one plant is required to start a new population.

Vegetative Propagation in Weeds

- Offspring are genetically identical to the maternal parent.
- Within population diversity can be hard to predict.
    - Is it a mixture of clones?
    - Is it only one clone?
    - Mutations are preserved.
- Among population diversity also can be hard to predict.
- It only takes one plant to start a new population.

Mixed Mating System in Weeds

- May be the most common method in weeds.
- The distribution of heterozygotes, homozygotes, and clones depends on the proportion in the original population.
- Some common combinations:
    - Facultative selfing in an outcrossing species.
    - Predominately selfing with a low level of outcrossing.
    - Outcrossing with clonal capacity.
    - Facultative apomixes.

Apomixis

- An asexual form of reproduction where a seed is formed without fertilization.
    - Somatic cells or gametes in the mother plant develop into an embryo.
    - Examples - Hawkweed (Hieracium spp.), dandelion (Taraxacum), some citrus (Citrus), blackberries (Rubus spp.), and Kentucky bluegrass (Poa pratensis).

Advantages of Seed Production

- Much more genetic diversity.
- Seed are a complete package.
    - Preserved in time.
    - Preserved against harsh environments.
    - Seed are easily dispersed.
    - Many seed are produced.
    - Depends on the presence of competitors.

Examples of Seed Production

Weed species	Seed per plant	Competitors
Without competitors		
Powell amaranth	100,000	No crop
Scentless chamomile	Up to 300,000	None
Common purslane	Up to 2,000,000	None
Common chickweed	30,000	No crop
Wild mustard	30,000	No crop

With competitors		
Velvetleaf	60-70	Corn
Powell amaranth	30,000	Brocolli
Giant ragweed	230	Corn
Barnyardgrass	100,000	Tomato
Green foxtail	28,700	Soybean
Wild mustard	328	Wheat

Disadvantages to Seed Production

- Species that produce many reproductive units (seed) may be less competitive than those that produce primarily vegetative growth.
- Producing seed uses resources that could be used for plant competition.

Vegetative Reproduction

- Many types of vegetative propagules are known.
- Vegetative reproduction occurs through modification of a stem that grows horizontally.
    - Underground stems or rhizomes,
    - Roots,
    - Aboveground stems or stolons,
    - Corms, bulbs, and tubers (storage for carbohydrates).

Advantages to Vegetative Reproduction

- Can create clones of a successful genotype and provide genetic stability.
- Can establish a new population from one plant or propagule.
- Storage organs propagate rapidly and in large quantities.
- Usually difficult to control.

Disadvantages to Vegetative Reproduction

- Limited genetic variability and mutations are preserved.
    - Less adaptable.
    - Expected to be well controlled by biological agents.
- Relatively limited dispersal.
- Shorter lifespan of propagules compared to seed.

- Sensitive to temperature and moisture extremes.

Management of Seed Bearing Weeds

- The juvenile period in weeds is often very short:
  - May produce seed quickly.
  - Must be aware of seedlings and control them before they set seed.
- Weed seed may be viable even if immature.
  - You should remove the entire plant from the field if pulled.
  - You should apply weed control before seed are viable.
  - Sometimes a late application of a herbicide can reduce viablity.

Weeds that Reproduce Vegetatively

- Good candidate for herbicides.
  - Phloem mobile herbicides translocate into underground structures and kill the entire plant.
- Good candidate for biological control.
  - The equilibrium that exists between populations of insects and weeds usually maintains the weeds at an acceptable threshold.
- Not good candidates for mowing, hand-pulling, and grazing of these weeds due to their extensive underground structures and carbohydrate reserves.
  - Can even increase due to the subsequent changes in apical dominance.

Reproduction of Perennial Weeds

Usually perennial weeds first infest an area by seed. Reproduction by seed is called sexual reproduction. It requires the fertilization of an egg by sperm, usually in the form of pollen. Pollination of the egg in a flower results in seed that is capable of producing a new plant.

Seed production varies greatly among and within weed species in part due to environmental variability between years, competition from neighboring plants, and genetic variability. For example, while Canada thistle has been observed to produce as few as 680 seeds per plant, curly dock often produces more than 30,000 seeds per plant.

Vegetative Reproduction

The ability to reproduce vegetatively is a unique characteristic that promotes the survival of a perennial species. In vegetative (asexual) reproduction, a new plant develops from a vegetative organ such as a stem, root, or leaf. Several modifications of these organs are common in perennial weeds, such as underground stems (rhizomes), above-ground stems (stolons), bulbs, corms, and tubers. Although vegetative structures generally do not survive as long in the soil as do seeds, very small structures can result in a new plant. Canada thistle, for example, can produce a new plant from as small as a 1⁄4-inch section of root.

Vegetative reproduction can be as prolific as seed production. Yellow nutsedge (Cyperus esculentus) has been reported to produce more than 1,900 new plants and more than 6,800 tubers in 1 year. Whitetop (Cardaria draba) has been observed to spread over an area greater than 10 feet in diameter and produce more than 450 shoots in its first year.

Yellow nutsedge (Cyperus esculentus).

Reproductive Strategies

Perennial plants are classified based on their reproductive strategy as simple or creeping perennials.

Common dandelion (Taraxacum officinale). Buckhorn plantain (Plantago lanceolata).

1. Simple perennials: Simple perennials spread by seed and have no natural way to spread

vegetatively. However, vegetative structures can produce new plants when cut or injured. Curly dock (Rumex crispus), for example, can generate a new plant from a very small piece of cut root.

Roots of simple perennials often are large and fleshy. Common dandelion (Taraxacum officinale), buckhorn plantain (Plantago lanceolata), and broadleaf plantain (Plantago major) are examples of simple perennials.

2. Creeping perennials: Creeping perennials reproduce by creeping roots, rhizomes, or stolons. They often also reproduce by seed, and therefore are the most difficult perennial weeds to manage. Canada thistle (Cirsium arvense), quackgrass (Elytrigia repens), johnsongrass (Sorghum halepense), bermudagrass, and field bindweed (Convolvulus arvensis) are examples of creeping perennials.

Broadleaf plantain (Plantago major).

Spread of Perennial Weeds

Seed Dispersal

Seeds have no way to move on their own, but they are excellent travelers. Their ability to spread and remain viable in the soil for years makes eradication nearly impossible. Weed seeds are dispersed by:

- Wind,
- Water,
- Animals, including humans,
- Machinery,
- Crop seed, animal feed, hay, and straw.

1. Wind: Many seeds are well adapted to wind travel. Cottony coverings and parachute-like structures allow seeds to float with the wind. Examples of wind-dispersed seeds include common milkweed (Asclepias syriaca), common dandelion, Canada thistle, and perennial sowthistle (Sonchus arvensis).

2. Water: Weed seed often moves with surface water runoff into irrigation water and ponds, where it is carried to other fields. Weeds growing in ditch banks along irrigation canals and ponds are the major source of weed seed contamination of irrigation water.

Weed seed often remains viable in water for several years, creating a "floating seedbank" and allowing weeds to disperse over large areas in moving water. Field bindweed seed, for example, remains over 50 percent viable after being submerged in water for more than 4 years. Some seeds have special adaptations that aid in water travel. The seedpod of curly dock, for example, is equipped with pontoons that carry the floating seed.

3. Animals: Several weed species produce seeds with barbs, hooks, spines, and rasps that cling to the fur of animals or to clothing and then can be dispersed long distances. Common cocklebur (Xanthium strumarium) produces seeds that adhere tightly to fur or cloth.

Weed seed often is ingested and passed through the digestive tracts of animals. Animal droppings provide an ideal nutrient and moisture environment for weed germination. While only a small percentage of the seed remains viable after exposure to an animal's digestive enzymes, keep in mind that a single weed might produce thousands of seeds.

4. Machinery: Weed seeds often are dispersed by tillage and harvesting equipment. Seeds move from field to field on the soil that sticks to tractor tires, and vegetative structures often travel on tillage and cultivation equipment. Disc-type cultivation equipment is less likely to drag vegetative plant parts than are shovels or sweeps.

5. Crop seed, animal feed, hay, and straw: Weeds probably are spread more commonly during the seeding of a new crop or in animal feed and bedding than by any other method. Seed labels often indicate a tiny percentage of weed seed, but consider this example. If a legume seed contains 0.001 percent dodder (a parasitic annual; Cuscuta campestris) seed by weight, there will be eight dodder seeds per pound of legume seed. If the legume is seeded at 20 lb/A, 160 dodder seeds will be sown per acre. Despite an extremely low dodder seed percentage by weight, the small size of the seed, combined with rapid early-season growth, could result in an infested legume field within a single season.

The Weed Seedbank and Seed Dormancy

Not only can weed seed and vegetative tissue travel great distances to infest new fields, but once in the soil, weed seed can remain viable for many years. In any given location, the weed seedbank contains a vast library of weed species and ecotypes that are adapted to a great range of environmental conditions and are ready to germinate given the proper signal. A Minnesota study reported that a square foot of soil, 6 inches deep, contained from 98 to 3,068 viable weed seeds. This represents between 4.3 million and 133 million viable seeds per acre. The amount of time that a seed is capable of producing a seedling, or its viability, varies with weed species. In the extreme, lotus (Nelumbo nucifera) seeds found in a Manchurian lakebed were viable after 1,000 years. More commonly, the annual plant jimsonweed (Datura stramonium) has over a 90 percent germination rate after 40 years in the soil.

Additionally, many weed seeds remain dormant in the soil until the conditions for germination and survival are appropriate for that particular seed. Dormancy is the seed's resting stage and is the primary method of weed seed dispersal in time. Some weed seeds have seed coats that are impermeable to water and/or oxygen or are mechanically resistant. Others contain immature embryos or have a waiting period (called after-ripening) that must be completed before the seed will germinate. Seed dormancy is affected by environmental conditions, including temperature, light, oxygen, and the presence of chemical inhibitors.

Seedling Establishment— The Perennial Advantage

Perennial weed seedlings are relatively easy to control with timely herbicides and cultivation. However, by 3 to 6 weeks after seed germination, most perennial weeds have begun to develop an extensive root structure that is capable of storing energy reserves. After this point, perennial weed control is a tedious process of depleting these stored energy reserves via a combination of tactics. Root growth is rapid in the first year after seed germination. Roots of Canada thistle can penetrate 3 to 6 feet deep in the soil in the first year. By 14 weeks after emergence, one johnsongrass plant was observed to produce more than 85 feet of rhizomes.

Canada thistle (Cirsium arvense).

Perennial Weed Interference with Crops

Competition for Resources

Weed and crop plants typically compete for three resources: water, nutrients, and light.

1. Water: Water often is the primary factor that limits crop production. Weeds often require more water than crops and are more efficient at capturing and using available moisture. Rooting depth and feeding area per plant often are greater for weeds than crops. For example, the annual barnyardgrass (Echinochloa crus-galli) requires more than 80 gallons of water to produce 1 pound of dry matter, compared to 60 gallons to produce 1 pound of wheat dry matter.

Quackgrass (Elytrigia repens).

2. Nutrients: Weed and crop plants compete for limited supplies of nitrogen, phosphorus, and potassium. Although the quantity of nutrients required by weeds and crops is similar at the same stage of growth, weeds are more successful in obtaining nutrients. As the superior competitor accumulates nutrients, it uses them to build larger, more expansive root systems to acquire additional nutrients and water, and to build a larger leaf canopy to capture light. Therefore, competition for a single resource has broad implications for the complex competition for other limiting resources. In general, the more competitive plant is the first to capture resources in quantities necessary for growth and to use them most efficiently.

Bermudagrass

3. Light: While nutrient and water resources can be captured by plants and stored for later use, light must be used as it is captured. Light competition is most severe when nutrients and water are supplied in excess, resulting in large leaf canopies. Plants that are good competitors for light have a leaf arrangement in which one leaf does not shade another and a tall or erect stem that allows them to intercept light over the canopy of neighboring plants.

Allelopathy

Allelopathy is a form of plant interference that occurs when one plant, through living or dead tissue, produces a chemical that interferes with the growth of another plant. Quackgrass, nutsedges, and johnsongrass are examples of perennial weeds that produce allelopathic chemicals that might inhibit crop growth. For example, yellow nutsedge reduces corn growth.

Chemical inhibitors can be produced and released during the growing season or released during plant decomposition. In an effect similar to herbicide carryover, these chemicals might remain active in the soil and suppress future crop germination and seedling emergence. Allellopathic chemicals often are produced in response to stress or competition from neighboring plants, and therefore add to the complexity of competition for limited resources.

Perennial Weed Management

Prevention

The most basic and effective of all methods to control perennial weeds is prevention. As discussed earlier, there are several means of weed seed dispersal, most of which can be prevented. Ensuring clean crop seed, animal feed, and hay is the most important measure in preventing seed dispersal. Other methods of prevention include cleaning field machinery and harvest equipment when moving between fields, proper long-term manure storage to reduce seed viability after passing through animals' digestive tracts, and maintenance of weed-free irrigation water.

Crop rotation can be another effective method to prevent the establishment of perennial weeds. The most effective crop rotations for this purpose include not only crops that compete well with perennial weeds, but also those that allow the use of herbicides to control perennial seedlings.

Mechanical Weed Control

Cultivation, when combined with other management tactics, can be used to control seedlings before energy-storing vegetative tissue has accumulated. Mechanical control no longer is effective after energy has been stored in underground vegetative tissue. In fact, cultivation of established perennials can spread weeds by cutting roots and moving them to new areas.

Perennial weeds are more common in reduced-tillage fields, where there is little soil disturbance to disrupt the development of below-ground storage organs. Once perennial weeds are established in reduced-tillage fields, cultivation is ineffective and might increase the spread of vegetative roots.

In pasture and forage crops, frequent mowing or cutting can prevent weed seed production and reduce the amount of energy stored in below-ground structures. Most important, maintenance of a vigorous crop stand through proper fertility and water management, seeding density, and variety selection will allow the competitive ability of the crop to suppress perennial weed growth. This simple "handsoff" approach requires little additional input or management, but can greatly reduce weed seed production and root growth.

Chemical Weed Control

Perennial weed control with herbicides must be repeated for 2 to 3 years and combined with other management tactics such as mowing. The key is to get the herbicide into the roots. Herbicide activity relies on foliar absorption and transport from the leaves to the root system. Young leaves move nutrients from the root in an upward, above-ground direction, while more mature leaves transport photosynthetic products to the root system for storage. Thus, the most effective herbicide activity occurs as the product is transported to the roots with the products of photosynthesis.

Field bindweed (Convolvulus arvensis).

Herbicides are most effective on perennial weeds in the early fall, when weeds are transporting energy to the roots before winter dormancy. Treatment just before and during flower bud initiation also is effective, as the herbicide will be carried with photosynthetic products to the roots. To ensure the presence of sufficient mature foliage, apply postemergent herbicides either 1 to 2 weeks before cultivation or mowing, or after weed regrowth is at least 8 inches tall.

Barnyardgrass (Echinochloa crus-galli).

Current herbicide recommendations for perennial weed control are included in the Pacific Northwest Weed Management Handbook. Before using any product, be sure to read the updated label provided by the manufacturer. No herbicide may be used on a crop or site for which it is not labeled, but the absence of a particular perennial weed on the label does not prevent the use of the herbicide on that weed.

Biological Weed Control

Biological control is the action of parasites, predators, or pathogens in maintaining an undesired organism's population at a lower average density than otherwise would occur. The goal is not to eradicate the pest, but to reduce the population density to a level below that which causes economic damage.

Biological control, when working ideally, is self-perpetuating and therefore economical after the initial release of the control agent. Once the organism is released, it maintains a population level appropriate to the amount of available food in the form of the undesired weed.

Biological control of perennial weeds has had limited success to date for several reasons:

- Successful survival of the control agent sometimes is a problem.
- Biological control agents can be used only if they feed strictly on the target weed. Unfortunately, potential biological control agents often cannot distinguish weedy species from their valuable relatives.
- Weeds must be controlled early in the growing season, prior to reproduction or crop yield reduction.

Biological control is a slow process, and results are not guaranteed. Therefore, it is used most appropriately as a component of an integrated weed management system that relies on multiple tactics for perennial weed control. For example, the fungus Concholiobolus lunatus kills barnyardgrass seedlings with fewer than two leaves, but growth of larger plants is only slowed and plants recover. However, when the fungus is combined with a sublethal dose of atrazine (a dose that injures but does not kill the barnyardgrass), larger barnyardgrass plants can be controlled better than when atrazine is used alone.

Integrated Weed Management

As the previous example demonstrates, management of perennial weeds is most successful when multiple tactics are employed, such as the combination of chemical, mechanical, and cultural control. The use of multiple modes of action enhances control and might prevent or delay resistance to a single control tactic, such as herbicides. Integrated weed management, when combined with prevention and control of weeds outside of crop production areas, provides the best long-term management of perennial weeds.

Weed Dispersal

Weed dispersal is simply the transporting of weeds from one place to another through several means and agents. When we see agent of weed dispersal it is generally assumed of anything that takes the seed of plant or crops from one destination to another for the sole purpose of propagation so many dispersal methods can have different names from different field of endeavour.

Human agent weed Dispersal.

Dispersal of Weeds

Weeds are mainly dispersed by gutter, animals, explosion and man, there are also times when wind erosion can also be a distinct method of disposing weeds.

Mechanical Method of Weed Dispersal

When speaking of mechanical methods of seed dispersal we are talking about the various ways through mechanical means like motor vehicles caterpillars and moving of heavy-duty machinery along Bush parts that resource and the dispersal of weeds and their seed People spread many weeds around through their machines.

For examples of weed and seed dispersal method: Motor vehicles such as (automobiles and trucks) raise gusts of wind as they travel up and down highways. As a result, small small seeds or seeds with tufts of hair are easily carried by the wind and blown along the motor paths and highways with each passing motor vehicle.

In this way, many weeds are distributed along roadsides. Similarly, heavy-duty vehicles and their freight cars generate even stronger gusts of winds as they travel up and down railroad tracks. here are typical examples of weeds and seeds that are transported or dispersed which is moving them from one place to another through mechanical methods of weed dispersal. As a result, weeds having tiny seeds or seeds with tufts of hair are distributed up and down railroads. For example, Dwarf Snapdragon.

Wind Dispersal of Weeds

The structures of some weed seeds enable their distribution by wind. Such features as being winged, having parachute make them easily carried about by wind. Examples of weeds dispersed by wind and Tridax procumbens and Ageratum conyzoides (Goat Weed).

- Agrostemma,
- Nemophila,

- Dianthus,
- Poppy,
- Columbine,
- Thistle,
- Felicia,
- Bulrush.

Water Dispersal of Seeds

When we talk about water dispersal methods of weed we are invariably talking about running Waters after a heavy downpour or through irrigation method in the farm majorly through surface irrigation water as an important agent to human activity serves as one of the key ways by which seed or weeds on the farm are dispersed from one location to another Weed seeds can be carried by moving water such as surface run-oft, streams, irrigation and drainage water. They have adaptation lenluus (hat enable movement by water such as membranous sacs containing air of corky protuberances. These make the seeds buoyant on water so that they can float. They can also stay in water for a period without losing viability. Example of weed dispersed by water is pig weed.

Animal Dispersal of Weeds

Animals also help in weed dispersal. For instance, weeds that have hooks can be distributed by animals with hairy body because such weeds can attach themselves to the hair of the animals. Example of weeds dispersed by animals is Erogratis tenella. Also seeds of some weeds pass through the digestive tracts of animals, particularly birds without losing viability. On defecation, these seeds can germinate into new plants.

Man as an Agent of Weed Dispersal

Man can transport weed seeds over long distances as a result of movements of agricultural seeds, foodstuff, nursery stock and others. Some weeds or their seeds could cling to the wears of the farmers and these are deposited elsewhere. Good examples are Boerhevia diffusa (Pig weed) and Acathespermum hispidium.

Man disperses numerous weed seeds and fruits with raw agricultural produce. Weeds mature at the same time and height along with crop. Due to their similar size and shape as that of crop seed, man unknowingly harvest the weeds also, and aids in dispersal of weed seeds. Such weeds are called "satellite weeds".

Self-Explosion or Explosive Mechanism

This is brought about by the forces set up in dry walls of weeds fruits which cause explosion. This scatters the seeds in the fruits away from the parents plants. Such weeds are said to be self-dispersed. Examples are Sida acuta and Desmodium.

Many weeds have seeds or seedpods that cling to the fur or body of animals and the clothing of humans as a result of sticky hairs, hooked spines, or barbed awns. Examples of weeds using this method of dispersion include:

- Lappa minor (lesser burdock),
- Torilis arvensis (common hedge parsley),
- Bidens frondosa (common beggar's ticks).

Several other weeds have seeds that become sticky while wet, these enables them to cling to the bottom of hooves, webbed feet, or shoes. Types of weeds using this method of seed dispersal include plantago lanceolata (english plantain) and several chamaesyce spp. (prostrate spurges).

Weed Dispersal by Machinery

Weed seeds often are dispersed by tillage and harvesting equipment. Seeds move from field to field on the soil that sticks to tractor tires, and vegetative structures often travel on tillage and cultivation equipment and latter dropping them in other fields to start new infestation. Disc-type cultivation equipment is less likely to drag vegetative plant parts than are shovels or sweeps.

Some Common Garden Weeds, Common Names and Botanical Names

- Guinea grass Panicumm maximum,
- Elephant grass Pennisetum purpureun,
- Tridax Tridax prociimbem,
- Carpet grass Axonopus compressus,
- Goose grass or Africana Cynodon dactylon,
- Siam weed Eluesine indica,
- African or bush marigold Chromoleana odorata,
- Pig weed Aspilia Africana,
- Goat weed or blue top Boerhavia diffusa,
- Water leaf Ageratum conyzoides,
- Soft weed Talinum triagulare,
- Wild green Erogratis tenella,
- Tropical kudzu Amaranthus spinosus,
- Centro Calapogonium mucunoides,
- Blue flower or water grass Centrosema pubescens,

- Spear grass Commelina spp,
- Striga Imperata cylindrical,
- Wild potato or moring glory Striga senegalensis,
- Broom weed Ipomea spp,
- Emilia Sida acuta,
- Giant star grass Emilia sonchifolia,
- Stubborn grass Cynodon nlemfuensis,
- Sensitive plant Sporobolus pyramidalis,
- Bur weed Mimosa pudica,
- Acanthospermum hisidum.

Weed Seed Germination

Weeds that exist with crops early in the season are less detrimental than weeds that compete with the crop later in the growing season, and this principle has supported the timely use of weed management practices. Either early- or late-emerging weeds produce great proportions of viable seeds that can remain in the soil profiles for a long time period, contributing to the perpetuation and the success of weeds. As a result, in most arable crop systems, weed management strategies focus mainly on reducing weed density in the early stages of crop growth. However, confining weed management to a narrow temporal window increases the risk of unsatisfactory weed management due to un-favorable weather. Weed seed banks are the primary source of persistent weed infestations in agricultural fields and if their deposits are increased, greater herbicide doses are required to control weeds afterwards. Annual weed species increase their populations via seed production exclusively, whereas seed production is also important for the spread of perennials.

Consequently, it is preferable to focus on depleting the seed stock in the soil through time rather than viewing weeds just as an annual threat to agricultural production. This approach is reinforced not only by ecological but also by economic simulation models. False seedbed technique is a method providing weed seed bank depletion. The principle of flushing out germinable weed seeds before crop sowing forms the basis of the false seedbed technique in which soil cultivation may take place days or weeks before cropping. Germination of weed seeds is stimulated through soil cultivation. Irrigation is suggested to provide the adequate soil moisture required for sufficient weed emergence. In the case of false seedbed, emerged weeds are controlled by shallow tillage operations. Control of weeds and crop establishment should be delayed until the main flush of emergence has passed in order to deplete the seedbank in the surface layer of soil and reduce subsequent weed emergence.

False seedbed technique aims to reduce weed seed bank by exploiting seed germination biology.

Thus, the efficacy of such management practices is directly associated with all the factors affecting germination of weed seeds and seedling emergence. Soil temperature, diurnal temperature variation, soil moisture, light, nitrates concentration in the soil, and the gaseous environment of the soil can regulate seed germination and weed emergence. Except for the case of environmental factors, tillage is the most effective way to promote weed seed germination because the soil disturbance associated with tillage offers several cues to seedbank residents such as elevated and greater diurnal temperature, exposure to light, oxygen, and release of nitrates in the soil environment.

Impact of Soil Temperature and Water Potential on Weed Seed Germination and their Roles

The longevity of weed seeds into the soil profiles is attributed to the phenomenon of dormancy that prevents seed germination even when the environmental conditions are ideal. Dormancy is distinguished into two types: primary and secondary dormancy. The end of primary dormancy is sequenced by the establishment of secondary dormancy and this sequence has been defined as dormancy cycling. In adapted weed species, dormancy is alleviated during the season preceding the period with favorable conditions for seedling development and plant growth, while dormancy induction takes place in the period preceding the season with environmental conditions unsuitable for plant survival. Furthermore, seeds from summer annual species are released from dormancy by low winter temperatures. High summer temperatures may induce entrance of the same seeds into dormancy again, which is referred to as secondary dormancy. On the contrary, seeds from winter annuals are released from dormancy by high summer temperatures whereas low winter temperatures induce their entrance into secondary dormancy. Relatively dry seeds lose dormancy at a rate which is temperature-dependent. In hydrated seeds, high temperatures reinforce or induce dormancy whereas low temperatures between −1 and 15 °C may stimulate germination.

Timing of weed emergence is dependent on the timing and rate of seed germination, which is dependent not only on soil temperature and but also on moisture potential. Of the many environmental factors that regulate seed behavior under field conditions, soil temperature has a primary influence on seed dormancy and germination, affecting both the capacity for germination by regulating dormancy and the rate or speed of germination in non-dormant seeds. It has been recognized since at least 1860 that three cardinal temperatures (minimum, optimum, and maximum) describe the range of T over which seeds of a particular species can germinate. The minimum or base temperature (T_b) is the lowest T at which germination can occur, the optimum temperature (T_o) is the T at which germination is most rapid, and the maximum or ceiling temperature (T_c) is the highest T at which seeds can germinate. Seed germination rates also vary with increasing temperature as it increases in the suboptimal range and decreases above the optimum temperature.

To account for the effect of temperature on the progress of germination, the concept of thermal time has been developed. The application of thermal time theory to germination is based on the observation that for some species there is a temperature range over which the germination rate for a particular fraction of the seed population is linearly related to temperature. The base temperature Tb is estimated as the x-intercept of a linear regression of the germination rate with temperature. Once seeds have lost dormancy, their rate of germination shows a positive linear relationship between the base temperature and the optimum temperature and a negative linear relationship between the optimal temperature and the ceiling temperature. For the case of the summer annual

Polygonum aviculare (L.), Kruk and Benech–Arnold demonstrated that low winter temperatures alleviate dormancy, producing a widening of the thermal range permissive for germination as a consequence of a progressive decrease of the lower limit temperature for germination of the population (T_b). In contrast, high summer temperatures reinforce dormancy, which results in a narrowing of the thermal range permissive for germination through an increase of T_b.

Germination speed of Alopecurus myosuroides (Huds.) seeds decreased with temperature, whereas the final proportion of germinated seeds was not significantly influenced. Minimum temperature required for seed germination is different for various weed species. Minimum temperature required for seed germination has been estimated at 0 °C both for the winter annual A. myosuroides and the summer annual P. aviculare. However, Masin estimated the base temperature for Digitaria sanguinalis (L.), Setaria viridis (L.), P. Beauv. Setaria pumila (Poir.), Roem. & Schultes and Eleusine indica (L.), at 8.4, 6.1, 8.3, and 12.6 °C, respectively. Moreover, the mean Tb recorded for summer annuals Amaranthus albus (L), Amaranthus palmeri (S. Wats.), D. sanguinalis, Echinochloa crus-galli (L.) Beauv. Portulaca oleracea (L.), and Setaria glauca (L.) was ~40% higher as compared to the corresponding value recorded for winter annuals Hirschfeldia incana (L.) and Sonchus oleraceus (L.). Optimal temperature conditions required for terminating dormancy status vary among different species. For example, Panicum miliaceum (L.) seeds lost dormancy at 8 °C while P. aviculare seeds were released from dormancy at 17 °C. The two germination response characteristics, Tb and rate, influence a species' germination behavior in the field. Extended models should be developed to predict the effects of environment and agricultural practices on weed germination, weed emergence, and the dynamics of weed communities in the long term. This requires estimating the baseline temperature for germination for each weed species that are dominant in a cultivated area and recording seed germination in a wide range of temperatures.

The knowledge about seed germination for the dominant weed species of a cultivated area is vital for predicting weed seedlings emergence. The possibility of predicting seedling emergence is essential for improving weed management decisions. However, weed emergence is the result of two distinct processes, i.e. germination and pre-emergence growth of shoots and roots, which react differently to environmental factors and should therefore be studied and modeled separately. In temperate regions, soil temperature is probably the most distinct and recognizable factor governing emergence. Soil temperature can be used as a predictor of seedling emergence in crop growth models. Soil temperature can also be used for predicting weed emergence, but only if emergence can be represented by a simple continuous cumulative sigmoidal curve and the upper few centimeters of soil remain continuously moist.

Fluctuating temperatures belong to parameters that can remove the constraints for the seed germination of many weed species once the degree of dormancy is sufficiently low. In particular, the extent and number of diurnal soil temperature fluctuations can be critical in lessening seed dormancy of several species. For example, alternating temperatures at 25 °C increased germination of Amaranthus retroflexus (L.), Amaranthus spinosus (L.), and Amaranthus tuberculatus (L.) from 23 to 65, 8 to 77, and 9 to 57%, respectively, as compared to non-alternating temperatures. Fluctuating temperatures from 2.4 to 15 °C can terminate the dormancy situation in Chenopodium album (L.) seeds. Either four diurnal cycles of 12 °C amplitude or 12 diurnal cycles of 6 °C amplitude were necessary for the emergence of D. sanguinalis. The number of cycles of alternating temperatures needed to end the dormancy situation has to be investigated. In Sorghum halepense

(L.) Pers. a 50% increase in cycles of alternating temperatures can double the number of seeds that are released from dormancy. Furthermore, if the demand for fluctuating temperatures to terminate dormancy in the seeds of this species is not satisfied, a loss of sensitivity to fluctuating temperatures occurs in a proportion of the population. The variation among weed species regarding the demands for fluctuating temperatures for seed germination points out the need for further investigation regarding the effects of fluctuating temperatures in germination of noxious weed species in different regions around the world and under various soil and climatic conditions.

Soil moisture is a key parameter affecting the seed dormancy status of many species. First of all, the environmental conditions existing during seed development in parent plants and seed maturation affect the relative dormancy of the seeds. Less dormant seeds of Sinapis arvensis (L.) were produced from the mother plants under water stress conditions while similar results have been reported regarding either winter annual grass species Avena fatua (L.) or summer perennial S. halepense. Moreover, sufficient water potential has been noticed to increase the production of dormant A. myosuroides seeds.

The effects of water deficits on seed germination have been encapsulated in the "hydrotime" concept. This idea was first illustrated by Gummerson and further explained by. The model of accounted for dormancy loss during after-ripening through changes in the base water potential of the seeds' environment that permits 50% germination ($\Psi_{b(50)}$). Christensen confirmed that $\Psi_{b(50)}$ value of the population is decreased by the change in $\Psi_{b(50)}$ due to after-ripening. The $\Psi_{b(50)}$ value is saved as the $\Psi_{b(50)}$ value of the population and serves as the initial value for the next time step. The process continues until the $\Psi_{b(50)}$ value of fully after-ripened seeds is reached. The model described is only to consider dormancy changes, not only in relation to the thermal environment, but also as a function of the soil water status. The loss of primary dormancy does not secure some species germination if moisture demands are not met. For example, adequate water conditions are demanded to promote germination of Bromus tectorum (L.). Bauer assumed that the temperature-dependent after-ripening process in this winter annual occurs at soil water potentials below ~-4 MPa. Martinez–Ghersa reported that increased water content promoted seed germination of A. retroflexus, C. album, and E. cruss-galli.

The seed germination response to the soil water potential of wild plants could be correlated with the soil water status in their natural habitats. The models which aim to predict weed germination and emergence need to record seed germination in a wide range of water potentials. Seeds of various weed species require different values of water potential in order to germinate. For instance, the base water potential Ψ_b for A. myosuroides was estimated at −1.53 (MPa) in the study of Colbach whereas the corresponding value recorded for Ambrosia artemisiifolia (L.) was −0.8 (MPa) as observed by other scientists. The value of minimum water potential for the germination of S. viridis seeds was −0.7 (MPa) whereas the corresponding value recorded for Stellaria media (L.) Villars was −1.13 (MPa). Dorsainvil revealed that the base water potential for germination for Sinapis alba (L.) was at −1 (MPa). Regarding weed emergence, although seeds of many species can germinate in a wide range of water potentials, once germination has occurred the emerged seedlings are sensitive to dehydration, and irreversible cellular damage may occur. False seedbed is a technique that aims to deplete weed seed banks by eliminating the emerged weed seedlings. Thus, it is crucial to have knowledge about water demands for germination for the dominant weed species of the agricultural area where a false seedbed is planned to be formed. If these demands are not met, then they can be secured via adequate irrigation in the meantime between seedbed preparation and crop sowing.

The Possible Effects of Light, Gaseous Environment

The reaction of seeds to light signals is dependent on phytochromes that consist of a group of proteins acting as sensors to changes in light conditions. Cancellation of dormancy by light is mediated by the phytochromes. All phytochromes have two mutually photoconvertible forms: Pfr (considered the active form) with maximum absorption at 730 nm and Pr with maximum absorption at 660 nm. The photoconversion of phytochrome in the red light (R)-absorbing form (Pr) to the far red light (FR)-absorbing form (Pfr), has been identified as part of the germination induction mechanism in many plant species. Germination can be induced by Pfr/P as low as 10^{-4} and is usually saturated by <0.03 Pfr/Pr. The quality of light received by seeds may be more important than the quantity. There is evidence that Far-red light (FR, about 735 nm) can inhibit germination. Regarding the way weed emergence is influenced by light, given that FR or the ratio of FR to red light (R, about 645 nm) increases as plant canopies develop and solar elevation decreases with time after the summer period, emergence of sensitive species should be inhibited during the summer period. However, the practical significance of FR exposure for emergence in field settings is not well-known.

Solanum sarrachoides (Sendtn.) seeds germination rate was recorded over 90 % in both light and dark conditions. However, seeds of several species require light for germination. Lower than 16 % germination percentages were recorded for Anchusa arvensis (L.) M. Bieb. A. Fatua, and Lamium amplexicaule (L.) in light while germination of Matricaria perforata (Mérat), Galinsoga ciliata (Raf. Blake), and Sonchus asper (L.) Hill reached the level of 99% due to exposure to light. Germination of Bidens tripartita (L.), Carex flacca (Schreb.), Juncus conglomeratus (L.), and Scirpus sylvaticus (L.) was also found to be enhanced by light whereas with no exposure to light, seeds of Chenopodium bonus-henricus (L.) Rchb were not able to germinate (Khan and Karssen, 1980). However, the results of indicated that light was a requirement for germination in <20% of seeds within populations of Amaranthus spp. and also that the same seeds were the most dormant out of the total amount of seeds in the soil seedbank. Light and a higher temperature of 25 °C promoted faster, uniform germination of seeds of (Taraxacum officinale (L.), while in darkness, the achievement of 50% of the final germination percentage was delayed and a longer mean emergence time was needed. Other scientists noticed that germination of E. cruss-galli was increased by 65% under light conditions as compared to darkness. In contrast to the findings of the studies presented above, exposure to light has been found to have inhibiting effects on seed germination of Galium spurium (L.). In Bromus sterilis (L.) and some lines of A. fatua, germination is inhibited by a single R pulse and this inhibition is abolished by a subsequent FR pulse. Therefore, light appears in some cases to function as a dormancy breaking signal when deeply buried seeds are moved to shallower soil depths, whereas in other cases exposure of seeds to light may inhibit germination in various ways depending on the species and circumstances. For the purposes of emergence modeling, the role of light requirements for seed germination needs to be investigated as they set a large and unexplored research area.

There is evidence showing that other environmental factors, such as nitrates and gases, can also regulate seed bank dormancy. For instance, germination of Sysimbrium offcinale (L.) Scop. is dependent on the simultaneous presence of light and nitrates, while in the case of Arabidopsis thaliana (L.) Heynh. nitrates modify light-induced germination to some degree. The seeds of summer annual species, S.officinale, showed increased sensitivity to nitrates and lost dormancy during the winter season. Regarding the winter annual S. arvensis, Goudey recorded maximal germination frequencies when NO_3^- content ranged from 0.3 to 4.4 nmol seed^{-1} for applied NO_3^- concentrations between 2.5 and 20 mol m^{-3}. Germination was significantly lower in seeds containing more than 5 nmol NO_3^- . Although the mechanisms by which nitrates stimulate dormancy loss remain under investigation, they maybe act somewhere at the cell membrane environment. The evaluation of the effects of nitrates in regulating weed seeds' germination and weed emergence is an area of interest for weed scientists and research needs to be carried out to get a better knowledge regarding this issue. There is also evidence that the range of pH values can promote germination of important weed species. For instance, Pierce noticed that seed germination of D. sanguinalis decreased with increasing pH when soil was amended with $MgCO_3$, whereas maximum root dry weights occurred at ranges from pH 5.3–5.8. A pH range of 5–10 did not influence seed germination of E. indica. Cyperus esculentus (L.) germination rate at pH 3 was 14% as compared to 47% at pH 7, while germination of Sida spinosa (L.) was highest at pH 9. In the experiment by Eupatorium adenophorum (Spreng.) germinated in a narrow range of pH (5–7) whereas other researchers recorded a 19–36% germination rate for Conyza canadensis (L.) Cronquist. over a pH range from 4 to 10. As a consequence, another area available for research is the role of soil pH on seed germination and weed emergence especially in fields where false seedbed technique has been planned to be applied.

Oxygen and carbon dioxide are two of the most major biologically active gases in soil. Oxygen concentration in soil air does not usually fall below the limit of 19%. During storage of seeds in soil, oxygen can have both detrimental and beneficial effects on the dormancy status of weed seeds. The cycle of secondary dormancy in the seeds of A. fatua was attributed to hypoxia. Hypoxic conditions did also cause a decrease in the germination capacity and rate of Datura stramonium (L.). Moreover, B. tripartita seeds showed increased germination rates under 5 and 10% oxygen concentration as compared to the germination rate recorded under 21% oxygen concentration. Germination of E. crus-galli was increased with oxygen concentrations in the range among 2.5 and 5% and declined when the oxygen concentration level was above 5% citepbib20. However, low oxygen concentration or the inability to remove anaerobic fermentation products from the gaseous environment directly surrounding the seed may inhibit seed germination. The results of Corbineau and Côme indicated that low oxygen concentrations, or even hypoxia, can terminate dormancy situation in the seeds of Oldenlandia corymbosa (L.). The results of Experiment 1 carried out by Boyd and Van Acker revealed that oxygen concentration of 21% highest led to 31, 29, and 61% increased germination of Elymus repens (L.) Gould. as compared to oxygen concentrations of 5, 10, and 2.5%. In the same experiment, the greatest germination rate for Thlaspi arvense (L.) was also recorded with 21% oxygen concentration.

The levels of carbon dioxide in soil air ranges between 0.5 and 1%. When soils are flooded, the ratio of carbon dioxide to oxygen typically increases and can have detrimental effects on seed germination and seedling emergence. In very early studies, concentrations of carbon dioxide in the range of 0.5 and 1% have been reported to have a dormancy breaking effect in seeds of Trifolium subterraneum (L.) and Trigonella ornithopoides (L.) Lam. & DC. Elevated carbon dioxide concentrations combined with low oxygen concentrations may further strengthen the signal to germinate and promote germination below the surface during periods of high soil moisture content, and this hypothesis was supported by the results of. Ethylene, a gas with a well-known role as a growth regulator, is also present in the soil environment, with its usual value of the pressure ranging between 0.05 and 1.2 MPa. At these concentrations, it has break-dormancy effects on seeds of T. subterraneum, P. oleracea, C. album, and A. retroflexus. According to Katoh and Esashi, at low concentrations in the soil ethylene promotes germination in Xanthium pennsylvanicum (L.) and similar observations have been made regarding A.retroflexus. However, these are results of old studies and it should be noted that a newer study stated that the role of ethylene in governing seed germination and seedling emergence cannot be clearly explained. The findings of another study where strains of a bacterium were evaluated as stimulators of emergence for parasite weeds belonging to Striga spp. were interesting. The bacterium Pseudomonas syringae (Van Hall) pathovar glycinea synthesizes relatively large amounts of ethylene. Consequently, whether oxygen, carbon dioxide, and ethylene influences weed seeds' germination and seedlings emergence is not yet clarified since variation has been reported among gases' concentrations and various weed species.

Weed Identification

Identifying all the weeds on a farm is not easy, and it is usually not necessary. However, correctly identifying major weeds can be an important first step toward effective control. Two weed species can look very similar at certain growth stages, yet differ greatly in life cycle, modes of reproduction, effects on crops, and responses to control tactics. Scientific names are more precise, as each species has just one valid scientific name at any one time. They are also less descriptive and a little harder to pronounce.

Common names are more user-friendly but less precise. A given weed might have two or more common names. For example, common lambsquarters is also called fat-hen or white goosefoot. Some common names have been attached to two different weeds; for example, lambsquarters has also been called pigweed in some regions, and the term witchgrass has been applied both to the perennial quack grass and to an annual weed closely related to fall panicum. In order to minimize such confusion, weed scientists have adopted official common names, such as common lambsquarters for the species Chenopodium album, and redroot pigweed for Amaranthus retroflexus.

Botanists position each plant species in the enormous family tree of the plant kingdom, which illustrates current best estimates of each species' genetic and evolutionary relationships with other plant species. A genus (plural genera) is a group of closely related species that share many characteristics of appearance, growth habit, and genetic makeup. Latin names are usually in two words, denoting the genus (e.g. Amaranthus) and the species (e.g. retroflexus).

Some plant species have distinct variants called subspecies, for which the Latin name has three words. Subspecies may evolve naturally as a species adapts to different environments (as many weeds do), or may be developed through plant breeding (e.g. cabbage, broccoli, kohlrabi, cauliflower, collards, and kale are all subspecies of Brassica oleracea). Subspecies can cross-breed, a fact that is important to understand. For example, Japanese millet (Echinocloa crus-galli ssp.

frumentacea), a useful grain and cover crop, is a subspecies of barnyard grass (Echinocloa crus-galli), the world's third worst weed.

Plant Families

The next larger grouping above genus is family. Some of the better-known plant families in agriculture include the grass family (including cereal grains, corn, sorghum, millets, and pasture grasses); the legume family (including peas, beans, clovers, and alfalfa); the brassica family (including cabbage, broccoli, cauliflower, radishes, and mustards); and the nightshade family (including tomato, potato, pepper, and eggplant).

Knowing a weed's plant family is important, as many economically important crops have weedy relatives in the same family, which may harbor insect pests or pathogens of the crop, or in some cases cross-breed with the crop itself. Occasionally, the weedy relative can play a beneficial role, acting as a trap crop to divert pests from the cash crop, or supporting important natural enemies of the pests as well as the pests themselves.

Identifying Characteristics of Weeds

Identify weeds with the help of a good field guide, manual, or taxonomic key to the agricultural weeds in your region. Collect a representative specimen or several specimens (recommended), and examine them closely, including foliage, stem, flowers, roots, and other belowground parts.

Plants are identified by visible characteristics that remain roughly constant among all individuals within a species. These can include:

- Leaf shape, leaf margins, and venation (branching pattern of leaf veins).
- Leaf structure (simple or compound).
- Arrangement of leaves on the stem.
- Presence or absence of hairs on leaves or other parts of the plant.
- Flower structure, color, size.
- Inflorescence (arrangement of flowers or flower clusters on plant).
- Size, shape, structure, color, and arrangement of fruits and seeds.
- Roots, rhizomes, and other underground structures.
- Life cycle (annual, biennial, perennial).
- Habit of growth (erect, prostrate, climbing, etc).

Characteristics that are more variable within a species, yet can help identify the weed include:

- Plant height and lateral spread.
- Degree of branching, arrangement of branches on main stem.

- Leaf size.
- Leaf and stem coloration; stem hollowness.

Other distinctive characteristics that help identify some weeds include:

- Presence of spines, thorns, prickles, or stinging hairs.
- Milky juice or sap when stem or leaves are cut.
- Presence of a leaf sheath surrounding the stem at each node.
- Stems square in cross section.

The more defining characteristics you can observe, the faster and easier it is to get a positive ID on the weed. Useful tools for identifying weeds include a ruler or folding rule to measure plants and plant parts; a hand lens or magnifying glass for examining small plant parts or features; a trowel, spade, or digging fork for exhuming intact root systems and other underground structures; and a weed identification guide or manual.

Some basic weed identification tools.

Noting when and how a weed emerges and grows can aid in identification. Summer annuals are mostly frost-tender, usually emerge between the spring frost-free date and late summer, and die at the first fall frost. Winter annuals emerge any time between the end of summer and early the following spring, flower and set seed in spring or early summer, and usually dry up with the onset of hot weather. Thus a weed that is still thriving after a fall frost is almost certainly not a summer annual such as pigweed, purslane, or galinsoga, and a weed that is succulent and vegetative in July is probably not a winter annual like henbit or yellow rocket.

How a weed first emerges from the soil can give clues to whether it is coming up from a seed or from a rhizome or other perennial underground structure. Dig up the emerging weed to see if it is a true seedling (easy to dig up, few fine roots, remains of seed or seed coat may or may not still be visible), or sprouting from a perennial root or other structure (harder to dig up, attached to a larger root, rhizome, tuber, or fragment thereof).

The small weeds in this photo are true seedlings, having germinated over the past two weeks. The larger weeds with somewhat arrow-shaped leaves are shoots of hedge bindweed that have emerged from rhizome or rhizome fragments within the top foot or so of soil.

Identifying Weeds at an Early Age

Flowering specimens are easiest to identify to species, because they display the greatest number of defining traits. In fact, plant families are delineated to a large degree by flower structure. However, growers often want to identify major weeds in earlier growth stages, even as emerging seedlings, in order to determine best management strategies while the weeds are small and relatively easy to control. This can be challenging to say the least. If a particular species has you stumped, let a few individuals develop into mature, flowering plants, then identify them.

Some weed manuals include good photographs and descriptions of seedling characteristics that allow identification of the weed's plant family or genus, if not species. Seedling characteristics include:

- Presence or absence, number, size, and shape of the seed leaves (cotyledons) on the aboveground part of the emerged seedling.
- Shape, arrangement, and size of the first true leaves emerging above the cotyledons.
- Stem and leaf coloration.
- Presence or absence of hairs on leaves or stem.

Finding the Weed's Photo ID

Once you have a weed specimen in hand, use one of these methods to identify its species:

- Compare the weed with photographs and written descriptions of similar-looking weeds in a field guide or manual.
- Use a dichotomous key, which consists of a series of "either–or" pairs of categories of plant

characteristics, through which you gradually narrows possibilities down to one or a few species. At this point, the key provides detailed descriptions, usually with photos to verify the ID.

- Use an interactive key, usually available on line or on a computer CD-ROM, which allows you to start with the most readily observable characteristics of the specimen at hand to identify the species or arrive at a short list of possible species IDs. The interactive key is usually better than the dichotomous key for identifying a weed in a vegetative (non-flowering) stage of development.

The direct comparison method is especially useful when you want to verify the ID of a weed with which you are familiar, or for which you have narrowed it down to a short list of possibilities. A strong likeness of the specimen to a photo or diagram, and a close match with the manual's written description suggests a correct ID. A clear discrepancy in one or more defining characteristics indicates an incorrect ID—try again. This method can be quick and efficient when the list of possible species is reasonably short—if the manual is well organized, is written for your region, and includes the weed in question! Time-consuming pitfalls include random guessing, and trying to verify ID of a weed that the manual does not include.

The dichotomous key is the time-honored method by which botanists and agricultural scientists have identified weeds, native vegetation, or cultivated plants for the past century or more. The dichotomous key can provide a definitive ID when skillfully used. Starting at the beginning of the key, read each pair of characteristic descriptions or categories, and choose the one that best matches the specimen at hand. Each choice gives a reference number directing you to the next pair of characteristics to examine. It works much like a treasure hunt.

Dichotomous keys often use a lot of botanical jargon, so make sure the key has a good glossary of terms before buying the book. One disadvantage is that some of the dichotomies may refer to characteristics not shown by the specimen (e.g. flower color when the specimen is vegetative), or may be difficult to see. In this case, you will need to explore both sides of the dichotomy. Errors early in the process can send you on a lengthy "wild goose chase" until the error is discovered.

Weed Identification Challenges

Any of these methods is only as good as the weed manual, key, or database used, the quality of the specimens available, and the observation skills of the user. If the weed in question is not included in the key or guide being used, you can waste a lot of time searching for it in vain! It is important to choose a manual or database that is written for your region and includes all of the region's major agricultural weeds. If a particular weed cannot be found in the reference you are using, it could mean any of the following:

- The manual is not sufficiently comprehensive.
- The plant was not recognized as an economically important agricultural weed in the region when the manual was written.
- The weed is a new invader from another region, country or continent.

If you cannot identify an important or abundant weed species present on the farm:

- Try to narrow it down to plant family or genus.
- Watch it closely to determine life cycle and habit of growth.
- Collect and try to identify additional specimens at different seasons and growth stages.
- Bring fresh specimens to the county Extension office or a nearby university weed science department for help with identification.

Farmers who bring fresh specimens of new or unfamiliar weeds in to Extension or university departments for positive ID can help with early detection of a new exotic invader or changes in weed geographic ranges related to climate changes or other factors.

Weed Ecology

"Weed ecology" is defined as the study of the interaction or relationship between a weed and another plant and its environment. In the past, the understanding of weed ecology has been a weak link in weed management, when in fact the ecology of a weed species could be the most important tool to determining the correct course of action. Some basic aspects of weed ecology include weed characteristics and weed classification. Some characteristics of plants that support weediness include rapid seed germination, rapid growth, the ability to take up and utilize large amounts of nutrients, prolific seed production, seed characteristics that promote dispersal, seed dormancy mechanisms, continual flushes of germination, the ability to adapt to various environmental conditions, and high tolerance to stresses. Furthermore, the classification of weeds can be comprised of population dynamic factors (i.e. habitat, growth form, life cycle, reproduction) and seed type. Once the basic ecology of a weed is known, proper management strategies can be implemented to prevent the plant from producing seed. However, weed ecology also encompasses seed production, the amount of seed produced and in what form, and seed dispersal through abiotic and biotic factors. Weed ecology is directly correlated with the plant community composition, the evolution of weeds (potentially through factors such as herbicide resistance), allelopathy, and competition. The importance of weed ecology to both natural and agricultural systems cannot be stressed enough.

Weed Diversity

The number of weed species that are typically found in conventionally managed crop fields is now a fraction of the levels recorded in the 1950-1970s, owing to increased fertiliser and herbicide use, simpler rotations and loss of field boundaries and semi-natural features in the landscape. For some, this loss of infield biodiversity is a concern, reflecting an erosion of the natural capital and ecosystem services on which sustainable production is founded. However, for others, it is seen as a measure of successful weed control and the concept of conserving weeds within cropped fields is, at best, incomprehensible and, at worst, an insult to the efforts of weed scientists over the past half-century to reduce the serious yield losses inflicted by weeds. This divergence in peoples' perception of weeds represents a philosophical fault line running through the weed science community and often reflects differences in scientific background (ecology vs. agronomy). Here, we argue that, rather than perpetuating this dichotomy, we should recognise that the objectives of maintaining arable biodiversity and preventing cropping systems becoming dominated by a few highly competitive, herbicide-resistant weeds both rely on a common set of ecological and management principles that should unite our research.

Two important points of clarification are needed to frame our argument. Firstly, the focus here is on weed species richness and evenness, not density or total biomass. There may be a benefit to increasing the total amount of plant resources for higher trophic groups on farmland (including seeds for farmland birds and pollen and nectar for pollinators and biocontrol agents), but the argument that weed biomass should be managed in crops specifically to provide these 'ecosystem services' is not our primary focus. Secondly, our emphasis is on the common weed flora rather than those rare and threatened arable specialists that have suffered the steepest declines in regional populations over recent decades and require specific conservation measures.

A More Diverse Weed Community is Less Competitive

Ecological niche theory argues that phenotypic differentiation between species results in contrasting ability to capture the resources required for growth and that co-existence is supported by spatial and temporal heterogeneity in resource availability. While selective herbicides have been a major driver of recent declines in weed diversity, the homogenisation of habitats through the use of inorganic fertilisers and simplified crop rotations has also selected for fewer dominant species with similar resource requirements to the crop. Although persistent seedbanks will continue to buffer the negative impacts of management, it is likely that the remaining diversity in conventionally managed crops is now driven to a much larger extent than in the past by repeated recolonisation from field edges or neighbouring ruderal habitats. This may explain the large numbers of 'chance' records observed in modern weed surveys and why diversity is higher in complex landscapes with smaller fields.

One consequence of a reducing niche breadth for weeds is that interspecific competition with the crop intensifies as the diversity of available resource pools decreases – the so-called Resource Pool Diversity Hypothesis (RPDH). In unmanaged systems, the supply and imbalance of limiting resources have also been postulated as the direct proximate cause of variation in species richness. Where increased weed diversity is an emergent property of spatio-temporal heterogeneity in resource supply, therefore, we would expect decreased crop competition. Here, we provide further support for the RPDH using data from the Broadbalk long-term experiment at Rothamsted. The Broadbalk experiment includes herbicide-free plots with contrasting fertiliser treatments and naturally assembled weed communities that we have compared with equivalent weed-free plots to calculate yield loss. A classic 'hump-backed' relationship of weed species richness is observed along the fertility gradient with the highest species richness observed on intermediate fertility plots where soil resources are most evenly balanced. When the data were grouped on the basis of weed species richness, a strong negative relationship with crop yield loss was observed.

The concept of habitat heterogeneity can be extended beyond fine-scale spatial patchiness of soil resources to capture a range of other management interventions that act as filters on the local weed species pool; examples include crop sowing date and intensity of cultivation. In functional terms, a greater diversity of crop types, nutrient inputs and cultivation practices will lead to a greater breadth of weed 'response' traits, consequently reducing the dominance of competitive 'effect traits' that impact crop yield. Robust evidence supporting this hypothesis of weed competition based on functional traits is currently lacking in the literature. In providing this evidence, long-term cropping experiments that result in naturally assembled gradients of weed functional diversity are preferable to short-term studies on artificially selected weed mixtures. Additional, trait-based, analysis of existing experiments such as Broadbalk and the establishment of new long-term cropping system experiments that focus on weed diversity and crop yield loss should, therefore, be a key research aim in the future.

A More Diverse Weed Community is an Indicator of Agronomic and Environmental Sustainability

The simplification of cropping systems and increased inputs of agro-chemicals have led to the dominance of a few competitive, highly adapted and widely distributed weed species. This is exemplified by Alopecurus myosuroides Huds. in north-west Europe, Amaranthus palmeri S. Watson in southern and central United States and Lolium rigidum Gaudin in Australia. As populations of these species increase, growers become more reliant on fewer herbicides, selection pressure increases and dominant weeds quickly adapt to new management including becoming resistant to herbicides. Many of the world's cropping systems are now afflicted by herbicide resistance, such that as weed diversity has declined, in many cases weed biomass has not; this challenges the ongoing sustainability of the whole system. Herbicide resistance is just one of multiple factors currently threatening the agronomic and environmental sustainability of modern agriculture. Other pressures on the system include declining soil health, pollution of water courses, greenhouse gas emissions and declining functional biodiversity (including pollinator populations). The loss of agro-ecosystem diversity has contributed to these problems; the cultivation of a narrow range of functionally similar crops on large contiguous areas is predicated on highly mechanised systems with high inputs of inorganic fertiliser and pesticides, leading to open, leaky systems with minimal organic inputs to the soil.

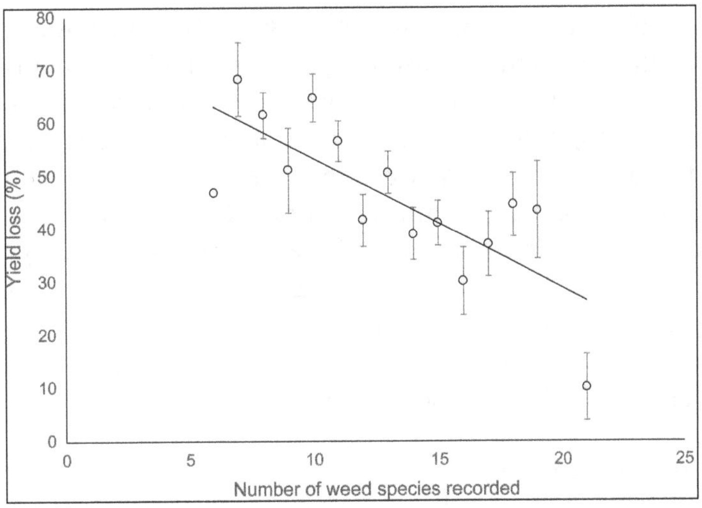

Relationship between weed species richness and crop yield loss on the Broadbalk winter wheat experiment begun in 1843. Weed species richness is assessed on herbicide-free plots annually, and weed diversity varies in response to contrasting fertiliser treatments. Winter wheat yield loss from weed competition can be calculated as a percentage of the equivalent plots with the same fertiliser treatments but where weeds are controlled with herbicides. Data are presented for 19 years collected between 1991 and 2014 (the plots were fallowed in some years during this period) and sorted by weed species richness. For each level of weed species richness, the average yield loss is presented with error bars indicated the standard error of the mean ($r^2 = 0.59$, $P < 0.001$). Plots with no nitrogen but with added phosphorus and potassium are excluded from the analysis as the abundance of leguminous weeds leads to facilitation and greater yield in the weedy plots compared to the weed-free plots.

In response to these stresses, there is a call for the 'sustainable intensification' (SI) of agriculture that increases food production without further adverse environmental impacts. This is a multifaceted and complex challenge that will require trading off multiple criteria using a range of metric. As well as being a direct threat to agronomic sustainability, we hypothesise that recent declines in weed species richness are correlated with the wider loss of cropping system resilience and that weed communities may represent a useful proxy for agronomic and environmental sustainability at the field, farm and landscape scale. There is potential, therefore, to develop an indicator of sustainable intensification built on the taxonomic or functional diversity of the weed seedbank that represents a legacy of previous management across the whole cropping system. Our hypothesis is that fields with low seedbank diversity have a less sustainable cropping system, both agronomically and environmentally, than a field with a more diverse weed community. Challenging this thesis will require a research effort to establish relationships between weed diversity and other metrics of SI, including soil health and functional biodiversity and the development of protocols that identify the appropriate measure of weed diversity. Because the emerged weed community in any given year is determined by the management specific to the crop being grown, as well as stochastic processes, sporadic assessments of the above-ground flora are not appropriate for this purpose. Rather, continuous assessments over the whole cropping system or sampling of the seedbank are required to capture the response of the weed flora to the range of management practices applied across the whole cropping system.

In conclusion, as weed biologists working at a single institution but whose research focuses on environmental and production endpoints respectively, we are convinced that the loss of weed diversity and the escalation of resistance to herbicides are mediated by an identical underlying cause: the simplification of agroecosystems and their associated weed management strategies. Given this, we propose that the goals of designing weed management systems that maximise production and maintain ecosystem functioning are entirely compatible and mutually reinforcing. We would, therefore, echo the call made by previous authors for weed scientists to integrate their work within the transdisciplinary framework that is required to meet the challenge of sustainable intensification and the transformation of cropping systems. In so doing, we would move weed science from being a parochial discipline towards an integral part of a broader research effort focussed on transforming the current, flawed paradigm of modern intensive agriculture.

Weeds in a Changing Climate

To sustain food production for the world's burgeoning human population, there is an urgent need to discover vulnerabilities and adaptive measures in managed ecosystems. It is unequivocal that food security, be either availability, accessibility, utilization, and/or system stability, is dependent on climate. Food security is potentially vulnerable to climate change since climate plays a pivotal role in determining growth, development, and perpetuation of all organisms. Climate is defined as the sum of weather conditions of a given area, quantified as long-term statistics of meteorological variables. These variables include temperature, wind, precipitation, and sunshine hours, all of which are essential for growth, development, and productivity of vegetation and in turn, human welfare.

In recent decades, changes in climate have caused significant impacts on natural and human ecosystems1. These impacts of climate change, irrespective of their cause, illustrate the sensitivity of natural as well as human ecosystems to variations in the function of climatic systems, interaction between its components, or changes in external forces either naturally or due to anthropogenic reasons. Of particular interest here is that agriculture may be jeopardized by climate change, since changes in weather factors have a significant effect on growth of all plant species, including crops and weeds. Rising atmospheric CO_2 and temperature are expected to pose both direct and indirect consequences for agricultural production, sustainability, water availability and, therefore, food security. However, in many ways, extremes of weather events associated with climate change are a more serious concern from farmers' perspectives on crop management as compared with more subtle changes brought by the actual increases in temperature, CO_2 levels, water availability and associated weather events. To cope with these extreme changes, future development needs to make adjustments in technology, management practices, and legislation.

At the more subtle level, it is recognized that weeds are aggressive, troublesome, and competitive elements within croplands. Contrary to other pests, weeds share a similar trophic level with crop plants, and by competing for scarce resources they cause enormous crop yield losses. The focus of this paper is on the dynamics of weed-crop competition and how they are influenced by climate, since this has important regional and global implications for food production. For example, the incessant rains during the kharif season (June–September) in India have made weed management a challenge, particularly in soybean-based cropping systems. Abrupt changes in climatic variables are likely to result in stressed crop plants which are vulnerable to attack by insect pests and pathogens, and makes them less competitive against weeds. It is important to note here that this area is extremely complex, as is shown by the work of Chen and McCarl who found that higher temperatures increase pesticide cost variance for maize (Zea mays L.), potatoes (Solanum tuberosum L.), and wheat (Triticum aestivum L.), while decreasing it for soybeans (Glycine max (L.) Merr.) They also reported that rainfall was directly proportional to unit land pesticide usage costs for these crops in the USA.

With a lack of precise information on the effects of climate change on agricultural pests, understanding of this issue remains a major obstacle for remedial measures. The ecological, environmental, and economic costs of not understanding these interactions can be substantial.

Climate Change and Weeds

The current atmospheric burden of the two most important greenhouse gases (carbon dioxide and methane), are unprecedented and have emerged as the greatest ecological challenge of the 21st century. The impact of climate change on weedy vegetation may be manifested in the form of geographic range expansions (migration or introduction to new areas), alterations in species life cycles, and population dynamics. Migration of weeds will subsequently result in a differential structure and composition of weed communities within natural and managed ecosystems. Through the lens of climate change, researcher outlined three distinct types of shifts in weedy vegetation (range, niche, and trait shifts), occurring at different scales (landscape, community, and population scales), respectively. Changes in weed biology, ecology, and interference potential, in the wake of climate change, will result in complex crop-weed interactions that necessitate alternative adaptive mechanisms. There is a general perception that climate change will result in a differential growth pattern between crops and weeds, as major weeds of the world have the C4 pathway and they will become more competitive, although this is certainly not a simple matter due to the adaptive mechanisms in weedy species.

Weed Response to Increasing CO_2 Levels

There is an ever-growing consensus that the earth's climate is changing, and despite the efforts made to reduce CO_2 emissions, there is an increasing pressure to identify adaptive mechanisms in agro-ecosystems. The record of atmospheric CO_2 obtained from Mauna Loa observatory at Hawaii indicated a 20% increase from 311 ppm in the mid-1950s to 375 ppm in 2001 (Keeling and Whorf, 2004), even though Mauna Loa and other global monitoring sites are situated in areas well away from regions of rapid CO_2 production. Previous studies have quantified a difference of 80 ppm in the CO_2 concentration between urban and suburban areas. This observation suggests that although data from Mauna Loa observatory mirrors the global increase, regional increases may be even more substantial due to rapid urbanization and intensive cropping, especially in Asia. This increase will continue in the near future, with estimates suggesting that it may reach 600 ppm, while the Inter-Governmental Panel on Climate Change have suggested, as a conservative estimate, 700 ppm by the end of the century.

The projected increase in atmospheric CO_2 is known to favor net photosynthesis in C_3 plants by limiting the loss of CO_2 via photorespiration and increasing the CO_2 concentration gradient from air to the leaf interior. By contrast, plants with a C_4 photosynthetic pathway manifest little response to elevated CO_2 as they have an internal mechanism to concentrate CO_2 at the site of CO_2 carboxylation. Thus, from this perspective, ongoing increase in the atmospheric CO_2 concentration will have crucial implications for weed-crop competition and crop yield losses. Numerous studies have addressed weed-crop interactions by evaluating the comparative growth and physiology of C_3 crops and C_4 weeds, and concluded that an elevated CO_2 concentration generally favors the vegetative growth of C_3 plant species over those with C_4 pathways. However, not all crops are based on C_3 pathways, and not all weeds are C_4 based. Hence, while the above concept is relevant for C_3 cereals such as rice, which compete, in the main, with C_4 grassy and broad-leaved weeds, this situation is not universal. There are many C_4 crops of economic significance, such as maize, sugarcane, and sorghum, which have competition from important C_3 weeds, for example, Chenopodium album L. This implies that weed-related yield losses of C_4 crops will tend to increase under elevated CO_2, but this will not occur with C_3 crops, as elevated CO_2 will be a crucial factor in realizing the potential benefits of CO_2 fertilization.

Notwithstanding this understanding, the abundance and appearance of weeds varies according to regions, crops, and management systems, which complicates management approaches. For example, Phalaris minor Retz. a C_3 species, is a problematic weed in wheat in the Indo-Gangetic Plains of North India. The C_3 weeds in other areas include Avena fatua L. Chenopodium album L. Cirsium arvense (L.) Scop. Convolvulus arvensis L. and Ludwigia hyssopifolia (G. Don) Exell. Weedy rice (Oryza sativa L.) is also a C_3 weed in rice in many Asian countries, including Vietnam, Sri Lanka, Malaysia, Thailand, the USA, and the Philippines. The third important grain crop, maize, is a C_4 species, and as noted, although there were some variations in the experimental conditions, C_3 species generally responded more favorably than C_4 species to an increased concentration of CO_2. Of further interest is that researcher indicated a greater physiological plasticity and genetic diversity among weedy (red) rice relative to cultivated rice, which may impact weed-crop competition with increased atmospheric CO_2.

It is becoming clear that predicting competitive outcomes based on species grown in isolation, may not adequately quantify crop-weed competition as a function of increasing CO_2, as weeds usually occur in a mixture. Therefore, there is a need to evaluate the effects of weed competition on crops in an environment of mixed of weeds and crops since most of the studies on the effect of CO_2 on crops and weeds have included weed and crop species in isolation. Only a few studies have examined the response of crops and weeds to CO_2 in competitive environments and a very little attention has been given to the effect of elevated CO_2 on the geographical distribution of weeds in managed ecosystems. In a study by researcher, an increase in CO_2 concentration resulted in a significant decrease in plant relative seed yield of a cultivated rice (C_3 crop species) variety, but the reverse for a weedy rice biotype, also a C_3 species. It is thought that this was probably due to a greater physiological plasticity and genetic diversity between a wild and cultivated lines.

Researcher opined that the increase in atmospheric CO_2 might also change the biology of invasive weeds. Presumably, an increase in CO_2 concentration stimulated growth and development of many invasive plant species, for example Cirsium arvense L. an invasive perennial C_3 weed species, had registered a 70% increase in growth with elevated CO_2. Increasing CO_2 levels may also increase wind dispersal of weed seeds by either increasing the height of the weed plant or by increasing the plant size. Some of these wind-dispersed invasive weed species are Cirsium arvense. Sonchus arvensis L. Sonchus oleraceus L. and Carduus nutans L.

Changes in weed communities in response to crop establishment methods, wet or dry moisture conditions, tillage regimes, and other management practices are well established in the literature. Nevertheless, very few studies have focused on changes in weed communities in the backdrop of elevated CO_2. Variations in the weed competitiveness response to elevated CO_2 among diverse lines of rice may necessitate screening of vulnerable and resistant cultivars for wider adoption. Although rice could benefit from CO_2 fertilization, the greater response of its wild relatives, particularly the weedy biotypes of rice, could offset the associated benefits and competitive outcomes as crop yield losses could increase.

Herbicide efficacy may also decrease as CO_2 concentrations increase. An increase in CO_2 induces morpho-physiological and anatomical changes in plants affecting the rate of uptake and translocation of herbicides. C_3 plants showed a decrease in stomata number and conductance and increased leaf thickness, which might interfere with foliar uptake of herbicides, as well as an increase in starch accumulation on the leaf surface. Moreover, perennial weeds may become even more noxious, if vegetative growth is stimulated as a result of increased photosynthesis in response to elevated CO_2. These changes are expected to reduce the efficacy of the most commonly used herbicides, such as glyphosate, due to a dilution effect, although the precise mechanism conducive to increased tolerance to glyphosate remains elusive. This could be due to less translocation as the root system becomes vigorous. In addition, an increase in the root-shoot ratio may play a critical role in herbicide efficacy. CO_2-induced increase in root biomass could make perennial weeds harder to control in a higher CO_2 environment. Thus, research is needed to assess the comparative growth and physiological response of C_3 and C_4 weeds of different age groups (seedlings or mature plants), and their molecular and biochemical bases to herbicide tolerance under ambient and elevated CO_2 concentrations. Allocation of resources to below ground parts, source-sink relationships, and mitochondrial respiration also need to be reassessed in the wake of climate change scenarios.

Weed Response to Elevated Temperature

Atmospheric temperature is regarded as an important indicator of weed species distribution in a geographical area. Its rise could alter weed proliferation and competitive behavior in weedy vegetation as well as in crop stands. The indicated likely climate change may favor C_4 over C_3 weeds, since under conditions of elevated CO_2, reduced CO_2 solubility, and decreased affinity of RUBISCO for CO_2 would deter C_3 photosynthesis. As a result, a variation in weed distribution will affect the world's most important cropping systems, for example, rice-based cropping systems through weed shifts to high latitudes and altitudes. In addition, Striga spp. might extend their range to moderate climatic zones. However, Striga asiatica (L.) Kuntze is relatively insensitive to temperature, and changes in the geographical range of the host plants seem to play a critical role in its distribution rather than the direct effects of temperature. If this concept is generalized for all parasitic weeds in the Orobanchaceae, these weeds could pose a serious threat to global crop production, especially in fodders, in the near future. Many C_s weeds, such as Amaranthus retroflexus L. Setaria sp. Digitaria sp. Sorghum halepense (L.) Pers. and Paspalum dichotomiflorum (L.) Michx. are expected to expand further north, which would have a more pronounced effect in the northern Europe, where the number of weed species is lower than in the south. Milder and wetter winters would tend to increase the survival of winter annual weeds, while thermophile summer annuals will grow more profusely in areas with warmer summers under prolonged growing seasons, enabling them to grow further north.

Climate change would spread arable weed species. For example, Datura stramonium L. which needs high temperature for profuse growth, would become a more competitive candidate under the climate change scenarios. Warm temperature regimes augmented the abundance of Hieracium aurantiacum L. in Australia through accelerated growth and reproduction.

Under warmer conditions, Setaria viridis (L.) P. Beauv. germinated later in the (August) season. This was a beneficial temporal non-synchrony with emergence of a maize crop, avoiding crop-weed competition. In contrast, a recent study indicated that this species would be a problematic weed in maize-based cropping systems elsewhere, through synchrony with maize emergence, which is probably due to stimulation by increased temperature. Therefore, S. viridis would become a competitor of maize at enhanced temperatures at the time of emergence. This has implications for the northern part of the Central Europe, where temperatures are still below the optimum for this species. Similarly, Rottboelliia cochinchinensis (Lour.) Clayton could invade the central Midwest of the USA and California from Gulf Coast states, with a 3 °C rise in temperature. If new weeds are introduced into a non-native area, new and effective herbicides may be needed.

Increased temperature had a greater effect on plant phenological development than elevated CO_2. Increasing temperature by 4 °C advanced the emergence timing of C. album and S. viridis by 26 and 35 days, respectively, and flowering time by 50 and 31.5 days. Increased temperatures strongly affected the biomass accumulation by annual grass species during their reproductive phase as compared with the vegetative phase, and such effects are more pronounced in C_3 than C_4 plant species. However, the increased temperature was believed to offset potential benefits of elevated CO_2 by reducing the reproductive output. The uptake and translocation of herbicides in plants and their persistence in soil will also be affected by rising temperatures. In addition to these effects, temperature will also affect the rate of water absorption and movement, which affects the rate of leaf development, cuticle thickness, and stomatal number and their aperture, thus indirectly affecting herbicide selectivity and efficacy.

Weed Response to Variation in Rainfall and Drought Spells

A variation in rainfall pattern and increased aridity consistent with a warming climate, could alter weed distribution and their impact on crop production. In the near future, aridity is expected to increase in many agronomically important areas, since an anticipated increase in temperature (1–5 °C) is expected with each doubling of the atmospheric CO_2 level. Rising temperatures also causes greater evaporation, and global trends in rainfall variability suggest that the monsoon regions will become drier, leading to a 5–8% increase in drought-susceptible areas. Trends in future rainfall prediction are difficult to predict, except to forecast more erratic rainfall and consequently, drought and flood would become recurrent phenomena. Under such a scenario, the distribution and prevalence of weeds will be problematic in crop ecosystems, and in particular summer droughts will affect weed management in spring-sown crops. Researcher postulated that under prolonged drought spells, C_4 and parasitic weeds like S. hermonthica will thrive better. Under excess water environments, weeds such as Rhamphicarpa fistulosa (Hochst.) Benth. will be favored. A change in rainfall patterns would favor hydromorphic weeds while prolonged drought spells will benefit C_4 over C_3 weeds. Under rainfed or dryland environments, little or no rainfall will hamper adequate land preparation for wet season rice as a result of limited water availability for flooding, especially early in the season when the rice is most susceptible to weed competition. This will limit traditional weed management in flooded rice and necessitate the use of herbicides. Rice yield losses are expected to be higher under such circumstances. Researcher emphasized that drought-tolerant rice cultivars would be required to prevent water stress-induced yield losses and to increase rice competitiveness against weeds under rainfed conditions. A change from transplanting to direct seeding of rice, in relation to water saving in South Asia, has already resulted in increased weed competition and changed weed dynamics.

Competition of cotton with Abutilon theophrasti Medic. and Anoda cristata (L.) Schltdl. increased under drought conditions. A yield reduction due to Xanthium strumarium L. was more pronounced in well-watered soybeans compared with water-stressed soybeans. An increase in rainfall provided greater competition to wheat growth and yield against C. arvense. Weed competition had little effect on crops under water deficit conditions, as the potential crop yield was already reduced by water stress. This was confirmed by researcher, who showed that rice could not survive under water stress conditions. By contrast, Amaranthus spinosus and Leptochloa chinensis (L) Nees survived under water stress conditions and produced a significant number of tillers/branches and leaves even at the lowest soil water content.

Increased rainfall frequency and intensity will have an adverse effect on uptake, retention, and activity of soil-applied herbicides. Increased cuticle thickness and leaf pubescence in response to drought, will reduce herbicide entry into leaves. These attributes can also affect growth and recovery of crops and weeds following herbicide application. Increasing aridity and drought will increase herbicide volatilization, and, moreover, frequent rain showers will reduce the "rain safe periods" available for herbicide application in a given cropping system posing multidimensional challenges for weed management. An unprecedented increase in rainfall (either as a single rain event or annually) may promote leaching of soil-applied herbicides, and subsequent ground water contamination. A general conclusion that can be drawn from the above discussion is that an increase in rainfall would lead to additional weed pressure, thus increasing the herbicide costs and overall cost of production of major crops.

Weed Response to the Interactive Effects of Climatic Variables

Climate change causes extinctions and alters species distributions of flora and fauna, and exerts inescapable impacts on various antagonistic and mutualistic interactions among terrestrial species. As noted earlier, the conventional concept that CO_2 enrichment favors C_3 plant species over C_4 by stimulating net photosynthesis is modified by other associated climate variables affecting this (simple) response. The interactive effect of the CO_2 enrichment will affect weed-crop competition simultaneously or sequentially in a complex manner, quite differentially from its effect on the photosynthetic pathway alone. Past research on climate change has focused on manipulating the plant response to the CO_2 concentration and not on the associated increases in temperature or drought. These anticipated changes in temperature and moisture projected under changing climates have obvious implications for germination and the spatial and temporal emergence of weed seeds and seedlings, which require more holistic investigation.

For example, dormancy, which is considered one of the major constraints to weed emergence is expected to be broken earlier or sooner due to greater moisture availability and warmer temperatures. Dormancy cycles observed in some species are known to be regulated mainly by soil temperature in temperate environments where water is not seasonally restricted, irrespective of their CO_2 response. Researchers demonstrated that the combined effects of soil water and nutrient stress limited the response of C_3 plants to elevated CO_2, but not C_4 species. Researcher suggested that water availability is a crucial factor that mediates species and community responses to rising CO_2 concentrations. Information regarding the interactive effects of elevated CO_2 with sub-ambient temperatures in either C4 weeds or crops is scarce. Few studies are available that have examined the growth and reproductive response of C3 weeds to the combined effect of temperature and CO_2. Weeds like Senna obtusifolia and Anoda cristata exhibited a higher range of growth stimulation under elevated CO_2 and temperature, however, other weeds (Triticum repen L. and Abutilon theophrasti) did not respond similarly.

Moreover, differential enhancement of C_3 crops and weeds by elevated CO_2 at sub-optimal temperatures should receive attention. Researcher found that competitiveness of a C_3 crop species (rice) relative to a C_4 weed species (Echinochloa glabrescens Munro ex Hook.f.) could be enhanced by elevated CO_2 alone, but a simultaneous increase in CO_2 and temperature still favored the weed. Remarkably, the interactive effect of CO_2 with water availability has been exclusively studied for crop species with little emphasis placed on quantifying differences between crops and weeds of the same photosynthetic pathway. Researcher found that rising CO_2 levels favored the growth of both a C_3 crop (soybean) and C_4 weeds [Echinochloa crus-galli (L.) P.Beauv. Eleusine indica (L.) Gaertn. and Digitaria ciliaris (Retz.) Koeler)] by improving water-use efficiency under drought, although greater growth stimulation in the C_3 crop was expected. Studies reporting the interactive effects of rising CO_2 levels, drought, and weed-crop competition are not common. It could be speculated that if CO_2 decreases, the water requirement of C_4 weeds relative to C_3 crops, C_4 weeds could still potentially compete successfully with C_3 crops under a high CO_2/drought situation. Ironically, few studies have focused on how CO_2-induced changes in phenological development could be modified by other climatic factors such as water supply and/or temperature.

The opinion of Rosenzweig and Hillel that rising temperature and CO_2 levels could make crop plants less competitive with weeds, together with a similar prediction a decade later by researcher that weeds would benefit more than cash crops, were both found to be true. Amaranthus retroflexus produced more seeds in barley cropping, albeit the growth of barley as well as the weed was reduced in southern Finland.

At a sub-optimum nitrogen level, elevated CO_2 reduced the competitive ability of rice against E. crus-galli. Hence, an increase in atmospheric CO_2 will exacerbate rice yield losses under low soil nitrogen status, owing to C_4 weed competition. Systemic investigations are needed to appraise the interactive effects of key environmental variables on different weed species and communities under diverse ecosystems.

Several modeling studies do not account for the impacts of an increased climatic variability which is a necessity in weed-crop interactions under the climate change scenario. It is a general perception that the only role of the endogenous process has been emphasized in the weed population dynamics models, which ignores exogenous variables such as climate. The population dynamics of weeds are a function of ecological interactions within and between plant populations, nutrient and water limitation, rainfall, temperature and stochastic forces. Using a reproduction function (R-function), they concluded that interactions between endogenous and exogenous factors are important for management of weed and invasive plants and climate change mitigation. Since predicting the impact of a weed under cultivated conditions at local scale requires a process-based modeling approach integrating local environmental conditions with the differential responses of the crop and weeds, researcher have developed a simulation model for winter wheat and a competing weed Alopecurus myosuroides in UK.

Herbicide-Climate Interactions

Herbicide effectiveness is dependent on the local climate/microclimate, and herbicides are no exceptions, particularly foliage applied post-emergence. A rise in temperature will increase volatility of certain herbicides such as trifluralin, rendering it less efficient. In the 1970s, rapid volatilization of surface-applied alachlor, butachlor, and propachlor occurred from continuously moist soils exposed to a constant 21 °C. Temperatures (day/night 32/22 °C and 26/16 °C) were not as critical as that of relative humidity in influencing acifluorfen (diphenylether group) phytotoxicity on Xanthium strumarium L. and Ambrosia artemisiifolia L. However, temperature had a significant effect on the degradation of imazapyr (imidazolinone group), flumetsulam (sulfonanilide family), and thifensulfuron (sulfonylurea group) in soil. Glyphosate absorption is dependent on the atmospheric temperature, as evident from Desmodium tortuosum (Sw.) DC. a C_3 weed. An increase in temperature or relative humidity increased the efficacy of mesotrione on

X. strumarium and A. theophrastii three-fold. The efficacy of the herbicide pyrithiobac (pyrimidinylthiobenzoic acid group) on Amaranthus palmeri L. was reduced at temperatures outside the range of 20-34 °C. Researcher found that relative humidity had the most significant effect on the phytotoxic action of glufosinate-ammonium, since this is attributed to changes in cuticle hydration and droplet drying. Studies under controlled environmental chambers in Australia using varying night/day temperatures of 5/10, 15/20, and 20/25 °C showed that Raphanus raphanistrum L. grown under cooler temperatures of 5/10 °C, was poorly controlled with 1,200 g ai ha-1 of glufosinate. By comparision, 100% mortality was achieved under 15/20 and 20/25 °C for the same dose, suggesting enhanced efficacy of glufosinate under enhanced atmospheric temperature.

Implications of Climate Change for Weed-crop Interactions

Uncertainty in agricultural productivity under a climate change scenario can be the result of plant-plant interactions through direct effects of a change in temperature and atmospheric CO_2, or indirect effects at the system level through shifts in crop–weed interactions and other biotic stresses.

Shifts in Weed Abundance, Distribution and Competitive Balance

Under ambient conditions, water availability and temperature are the principal determinants of species distribution, but there is the recent addition to this list of CO_2 concentrations through the lens of climate change. The changing climate variables may either increase the distribution range of weed species in response to a change in atmospheric temperature, or allow some non-potent weeds to dominate weed abundance as crop-weed interactions may increasingly favor C_3 weeds. Other than geographical distribution, the projected climate change might impact their population biology, causing them to move to new areas lying at higher altitudes and latitudes. Such effects have been proposed for Striga sp. which are expected to extend their geographic range. Climate change will alter the distribution of plant species and overall functioning and productivity of ecosystems. For example, increased abundance of woody vines as consequences of rising CO_2 levels has been associated with increased tree mortality and reduced tree regeneration in forests throughout the world. Similarly, an increase in parasitic weeds would become a serious threat to productivity of rice and sorghum crops under rainfed agriculture. According to researcher, most of the troublesome C_3 and C_4 weeds of the arable land are limited to tropical and subtropical regions, primarily due to low temperatures at higher latitudes. Preliminary data showed an increased tolerance in many weeds to low temperatures under elevated CO_2, which suggests the possibility of polar-ward expansion for many weed species.

Species either have to adapt in situ to new climatic conditions or undergo shifts in their distribution to more favorable locals. Researcher proposed that if climate change forecasts are realized, damaging endemic weed species of major cropping systems might experience a significant transformation in their host range, besides an overall increase in the chance of invasion by exotic invasive weed species. Besides agronomic weeds, there are also certain non-native weeds whose introduction to new areas can pose ecological and environmental hazards. Several studies have demonstrated that such weeds often benefit from carbonaceous fertilization. It is believed that under a climate change scenario, these invasive plants would be able to extend their geographic range as well as spread to new areas, including currently agriculturally productive regions. An expansion in the geographic range proposed for weeds such as Lonicera semperviens L. and Pueraria lobata (Lour.) Merr. in the past has now become a reality. Range expansion of arable and invasive weeds in connection with climate change must be studied as an integral part of crop-weed interactions.

In a composite stand of weeds in a cropped field (C_3 and C_4 plants), dynamics in insurgence and shifts of the weed populations in favor of specific species is expected over time. Climate change is likely to trigger differential growth in crops and weeds and will have significant implications for weed management across crops and cropping systems. The abundance, competitive ability, and survival of perennial weeds are expected to be higher, since a rise in CO_2 stimulates tuber and rhizome growth. Climate change will result in a greater frequency of extreme weather events such as frequent droughts and cold spells, so that weeds with less phenotypic plasticity may experience population declines. Lack of rainfall and prolonged drought will limit growth of arable crops and pastures, resulting in a lack of vegetation cover and bare ground, thus allowing invasion by more resilient drought-tolerant weeds. Increasing CO_2 could alter the competitive balance in a weed-crop mixture through its effect on photosynthesis and stomatal physiology, which is linked with the competitive balance between crops and weeds in a cropping system. The range of growth stimulation in response to elevated CO_2 needs to be determined for both crops and weeds with contrasting carbon fixation pathways, growing in variable densities and species compositions. Under conditions of higher temperature and drought, C_4 weeds such as A. retroflexus tend to dominate C_3 crops (e.g. soybean). The infestation of P. minor is expected to worsen in wheat fields with CO_2 increase. Likewise, weedy rice will compete more strongly with cultivated rice. Exploring differential mechanisms and responses that govern the success of weeds to invade new areas/ cropping systems and their ability to utilize growth resources, will be helpful in understanding the implications of rising CO_2 levels on plant-plant interactions. This also requires characterizing their damage niche.

Effectiveness of Weed Management and Adoption of Best Agronomic Practices

Climate change will indirectly affect the adoption and success of weed management strategies. Looming water crises have been recognized as a major threat to agricultural productivity notably in irrigated rice with long-term consequences for regional and global food security. Water requirements of irrigated rice are approximately 2–3 times higher than for any other upland cereal. Aerobic rice is a potential water-use efficient production system, but a high weed infestation has threatened its sustainability, which demands efficient and cost-effective weed management techniques. Frequent drought spells and erratic rainfall will affect productivity and sustainability of upland and low land rice production systems. There will be a trade-off between water-use efficient rice production methods and weed management. In upland rice, drought tolerance will be needed not only to cope with water scarcity but also to safeguard production losses against weeds by maintaining or improving a competitive edge. In aerobic or dry-seeded rice, the switch over from transplanting in respect of water saving induces qualitative and quantitative changes in rice weed flora. The inherent size differential of transplanted rice seedlings in conjunction with flooded environments provided a distinct competitive advantage, i.e. an earlier growth and germination over a wide range of weed species that otherwise are quite problematic in aerobic rice. With a dwindling water supply and more severe drought spells, flooding will not be available as a potential weed management tool in the near future.

Hand weeding was 35% higher when the flooding regime was altered from permanent to temporary flooding. This means that farmers lacking alternate means and resources to combat weeds will suffer significant yield losses. Moreover, dry tillage practices, alternate wetting and drying regimes, and extended periods during which soil is not flooded, will result in the insurgence of non-native and difficult-to-control weeds. Under drought conditions, rice (C_3) is already a poor weed competitor and will be under greater pressure due to increased competition from C_4 weeds, which comprise the majority of weed flora infesting rice fields. In rainfed rice, a lack of rainfall early in wet seasons may compel farmers to adjust their timing of land preparation and subsequent planting. This might affect synchronization of rice sensitive growth periods with emergence and active growth period of troublesome weeds. Hence, it seems that strategies aimed at mitigating climate change effects on crop production like drought-tolerant rice germplasm and water saving rice cultivation, will also have implications for weed management.

Increasing interest in conservation agriculture has created a reliance on glyphosate for weed management and the continuous use of this herbicide may result in evolution of resistant biotypes of major weeds. In wheat, resource conservation technologies, such as no-till systems, have emerged as an important breakthrough. However, adoption of no-till approaches which are characterized by minimal soil disturbance, may affect the abundance and floristic composition of weeds. Hardy weeds, such as Rumex sp. are expected to be higher in zero-till wheat fields.

Increased tolerance to glyphosate under elevated CO_2 has been recorded for both agricultural and invasive weed species. These alarming findings revealed a sustained increase in photosynthesis and growth of perennial weeds such as Elymus repens (L.) Gould. With a concurrent decrease in herbicide efficacy and increased potential of invasion and competition. Differential tolerance to glyphosate exhibited by certain weeds under elevated CO_2 is also an issue. Whilst the response of weeds such as A. retroflexus was not affected by elevated CO_2, Chenopodium album and Cirsium arvensis manifested a significant glyphosate tolerance. A variable response to glyphosate was observed even for invasive grass species possessing the same carbon fixation pathway. Hence, it can be inferred that some weeds will be more problematic in the near future in glyphosate-tolerant crops or under conservation agriculture. Another difficulty will be the knockdown of perennial weeds if glyphosate efficacy is reduced due to climate change. An increase in rhizome and tuber growth, coupled with an increase in biomass, would cause a dilution effect on any herbicide application, causing an increase in weed control costs. Direct effects of climate change on plant physiology, anatomy, and morphology will indirectly affect herbicide efficacy by influencing uptake, translocation, and metabolism. Changes in physical environments, such as drought spells or prolonged rainy seasons, may limit the field conditions necessary for herbicide applications. Climate change will have implications for all dimensions of chemical weed management including application, spray drift, persistence, metabolism, and herbicide efficacy. This justifies diversifying current weed management tactics as well as the urgency of a sound knowledge regarding the ecology and biology of weeds in a changing climate.

After a catastrophic climatic event such as drought or flood, weeds will have a greater chance to colonize and invade disturbed habitats. Chemical control measures may become less effective due to a change in the external environment (drier and warmer conditions) or changes in anatomy, growth physiology, and phenology of the target weed flora. Asexual reproduction through below-ground parts is always conducive to spread, irrespective of water availability. Extremes of moisture availability, viz. flood as well as drought, hinder physical management methods such as hoeing, inter-cultivation, etc. It seems that growers will have to carefully synchronize the timing of their control measures with the weed life cycle since these will also respond to climate change. The opinion of Chandrasena that adaptive responses should be based on a better knowledge on how plant communities will respond to climate change rather than ad hoc responses is therefore valid in the current context.

Weed Seed Longevity

Weeds can be prolific seed producers which means if a single plant is allowed to go to seed, its impact may be felt for many years to come. One example: the broadleaf weed Palmer amaranth has been found to produce as many as a million seeds per plant.

To compound the problem, the seeds produced by many weed species can remain dormant — hidden away in the soil for extended periods of time — and then sprout when conditions are right. While seeds of most annual weedy grasses die after 2 or 3 years, scientists have discovered that some broadleaf weed seeds can remain viable for many decades.

Scientific Proof of Longevity

One of the most famous experiments illustrating the longevity of weed seeds dates back to 1879. Professor William Beal at Michigan Agricultural College set out to determine how long seeds could be buried in the soil and still remain able to germinate.

Beal started with freshly grown seeds from more than 20 common weed species and buried them in 20 sand-filled bottles about three feet below the soil surface. At five- to 10-year intervals throughout his career, he would dig up a bottle to see if the seeds it held would germinate.

When Beal retired, he passed the torch to fellow scientists at what now is Michigan State University. They've continued Beal's research, and what they've found is bad news for farmers and home gardeners alike. More than 130 years after Beal's experiment began, some of the seeds he planted are still able to germinate. The heartiest has been moth mullein, a flowering biennial weed from the figwort family.

Why some Weed Seeds are Long-lived

While some weed seeds will survive for long periods, many others will die within a few days or weeks if they aren't in an environment suitable for germination. Why the difference?

Seed coats are an important factor. The hardest coats provide the best protective barrier and promote longevity. These tough coats may also help to keep the seed hidden from predators by keeping it from giving off any odors that might be detected. Burial in the soil also enhances seed survival by inhibiting germination and protecting seeds from predators and environmental stresses.

Interestingly, scientists have discovered some weed species produce seeds that vary widely in size and dormancy periods. Examples include poison hemlock, jointed goat grass, twoscale saltbush and common cocklebur. By staggering germination throughout the growing season, the likelihood is greater that at least some plants will emerge to favorable conditions, enhancing overall survival.

What Causes Seeds to Germinate?

A number of factors can trigger weed seed germination, including changes in temperature, moisture, oxygen and light. Tillage is an especially common trigger. It exposes buried seeds to warming sunlight and aerates the soil.

There also are factors that can inhibit germination. The soil is a factor – _especially for smaller seeds. It serves to block the light signals the seeds need in order to determine they are close enough to the surface to germinate. Some small seeds can't emerge if they're more than 5 centimeters (about 2 inches) underground.

The same seed coat that protects seeds from destruction can also inhibit germination. These hard barriers can prevent the seed from getting the moisture or oxygen needed to germinate and can forcibly enclose the embryo. Physically scratching the seed coat to weaken it makes germination more likely.

Another factor that can impact germination is allelopathy, a biological phenomenon that involves one organism producing biochemicals that positively or negatively influence the growth, survival and reproduction of other organisms. For example, crops like rye, sorghum, rice and wheat release biochemicals that can suppress weed seed germination.

How Weed Seeds Travel

Weed seeds can be spread in feed grain, straw and hay and can be transported by birds, animals, water and wind. But they also can travel in other less expected ways. Seeds can stow away on farm machinery, off-road vehicles and cars. They can hitch a ride with container-grown ornamentals or travel the country in a package of wild bird feed. Seeds with burr-like surfaces can travel attached to your clothes or to animal fur. Earthworms are known to collect weed seeds and move them into their burrows. In fact, scientists say more than two-thirds of all giant ragweed seedlings emerge from earthworm burrows. Seeds consumed by livestock can often survive and can be transported in manure.

The award for longest-traveling weed seeds would likely go to horseweed. Tiny horseweed seeds have tufts that act like a parachute and help them travel vast distances by air. Specially equipped model airplanes have discovered horseweed seeds in the Earth's planetary boundary.

Reducing Weed Seeds in Soil

Whether your weed control method of choice is your hand, a hoe, a plow or an herbicide, it is important to control weeds that sprout before they mature and go to seed. If you are careful about weed management, over time you'll reduce the number of viable seeds in the soil and make life a bit easier. Here are some of the techniques that have been shown to help:

- Some growers attach a chaff cart to the back of their harvester to ensure seeds on weeds remaining in the field at harvest time aren't dispersed and left to sprout in subsequent growing seasons. It's a technique common in Australia.
- Australian grower Ray Harrington created a system that mills the chaff collected from the field to destroy weed seeds. Residues can then be dispersed directly onto the field so moisture and nutrients are conserved.
- Carabid beetles have been shown to reduce the number of weed seeds in the soil bank. They are voracious eaters and can consume large quantities of seed.
- Diversified cropping systems can result in a high microbial biomass in the soil that promotes the decay of weed seeds.
- Deep tillage can reduce germination by burying weed seeds deep beneath the soil. Unfortunately, future tillage may simply bring them to the surface again. One alternative is to superficially disturb the soil to cause seeds to germinate, and then treat the seedlings that emerge with herbicides.

What if a Weed Grows from Tubers?

Seeds are not the only source of new weeds. Many perennial weeds can reproduce from stems, roots, leaves and underground roots or tubers. Frequent removal of the shoots of perennial weeds will eventually starve and kill the underground tissues.

In some cases, herbicide applications at timely intervals after emergence can also help control underground tissues. You'll need to be persistent and will likely need to use a variety of control methods to reach your goal whether using tillage, physical removal or herbicides.

Weed Crop Competition

Competition is nothing but the struggle for existence and superiority. Competition exerts a powerful force in the plant community, irrespective of the mechanism, tending towards limitation or extinction of the weaker competitors. Competition is maximum when available resources for crop growth become limited.

Competition is a negative interaction where individuals make simultaneous demands that exceed limited resources and, while both suffer, one individual suffers less. So, crop weed competition indicates competition between crop and weed in a natural ecosystem in response to resources struggle for their existence and superiority.

Crop weed competition occurs in two broad aspects:

- Direct competition for nutrient, moisture, light and space.

- Indirect competition through exudation and production of allelopathic chemicals.

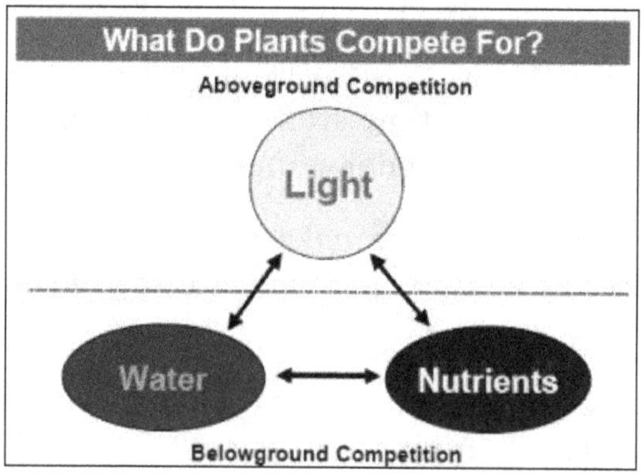

By and large, weeds appears much more adapted to agroecosystem than our present day crop plants.

Components of the Overall Competitive Effect

In an infested field it is possible to identify different components of the overall competitive effect:

- Intraspecific competition between plants of the cultivated species.
- Interspecific competition between plants of the cultivated species and weed species.
- Interspecific competition between plants of the different weed species.
- Intraspecific competition between plants of the same weed species.

Competition between weeds and crops is expressed by altered growth and development of both species. Interspecific competition occurs when two or more species coexist in time and space and simultaneously demand a limited resource. Intraspecific competition occurs when two or more plants of the same species coexist in time and space and simultaneously demand a limited resource.

Competition for Nutrients

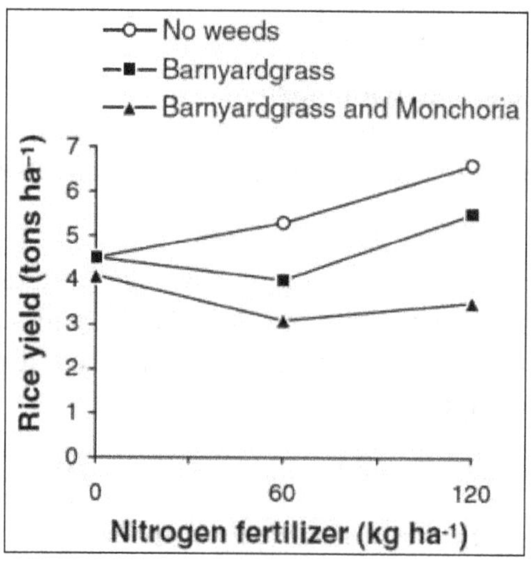

Plants compete mostly for nitrogen, phosphorus and potassium (but there are many others). Phosphorus is usually the most limited nutrient in aquatic ecosystems. Nitrogen is usually the most limited nutrient in terrestrial habitats. Potassium is often overlooked but some terrestrial weeds can grow well in K-rich soils. Approximately competition for nutrients constitutes an important aspect of weed crop competition. Weeds usually absorb mineral nutrients faster than many of our crop plants and accumulate them in their tissue in relatively large amounts.

Effect of nitrogen addition and two weeds (barnyardgrass and the annual broadleaf Monchorias) on rice yield.

Table: Kilograms of Nutrients Required to Produce Equal Amounts of Dry Matter.

Plant	Nitrogen	Phosphorus
Wheat	5.5	1.2
Lambrushquarter	7.6	1.6
Pig weed	5.1	1.4

- Species of Amaranthus, for example, often accumulate over 3% N in their dry matter.
- Chenopodium and Portulaca spp. are likewise potassium lovers with over 1.3% K_2O in their dry matter.
- Nutrient removal by weeds during the first 30 days of maize growth was 59 kg N, 10 kg P and 59 kg K per hectare, which was 7-10 times more than the nutrient removal by the crop.
- Weed possess not only a capacity for heavy nutrient absorption and accumulation but also gather tremendous quantities of dry matter.

Competition for Moisture

Competition for water occurs below ground between roots. The ability to absorb water is related to rooting volume. However, not only are the dimensions (breadth and depth) of rooting zones important: so is the degree of water extraction.

In general, for producing equal amounts of dry matter, weeds transpire more water than do most of our crop plants. In weedy fields, the soil moisture may be exhausted by the time the crop reaches the fruiting stage, which is often the peak. The consumptive use of water of a common weed. Chenopodium album as 550 mm against 479 mm for wheat crop itself. It is because weed can remove moisture from deeper depth of soil than crops.

Table: Water required to produce one pound of dry matter.

Plant	Litres of water
Wheat	227
Lambrushquarter	300
Corn	159
Pigweed	132-139

Competition for Light

Although it varies in duration, intensity, and quality, light regulates many aspects of plant growth and development. Neighboring plants may reduce light supply by direct interception: shading. Leaves are the site of light competition. Whenever a leaf is shaded by another, there is competition for light.

Light competition is most severe when there is high fertility and adequate moisture because plants grow vigorously and have larger foliar areas. Plants with large leaf area indices (LAI) have a competitive advantage with plants with smaller leaf areas.

Both light quality and quantity are important aspects of competition. Since the presence of dense leaf canopies reduces the quantity and quality of light available to weeds, competition for light is greatest when plant density is highest. Plant height defines an effective component of the competitive struggle for light. It becomes most important element of weed crop competition when moisture and nutrients are plentiful, and weeds have an edge over crop plants in respect of height. Light competition may commence very early in the crop season if a dense weed growth smoother the crop seedlings. Once a plant is shaded by another plant, increased light intensity cannot benefit it.

Critical Period of Weed Growth

Critical period of weed growth can be defined as that shortest time span in the ontogeny of crop growth when weeding with result in highest economic returns. The crop yield level obtained by weeding during this short span should provide crop yield sufficiently close to that obtained by the full crop season freedom from weeds. A fundamental principle of plant ecology is that early occupants on a soil tend to exclude the later ones.

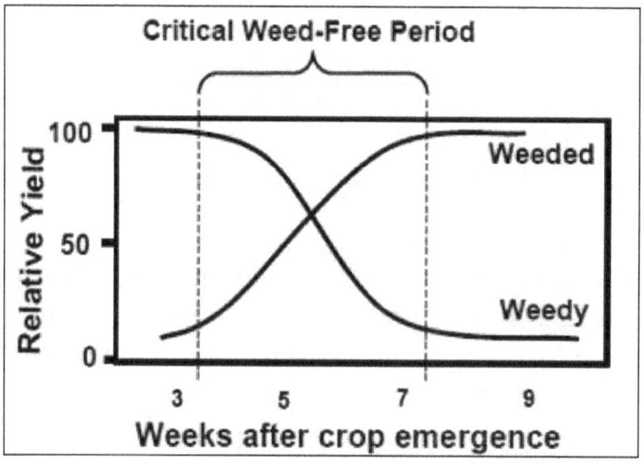

- On the basis of the plant ecology, crops required a weed free respite during the first one-forth to one-third of its growing period.
- Sharma found that in direct seeded rice, the critical weed competition period occurred 10-20 days after crop emergence. For the transplanted rice the critical periods of weed crop competition were identified. These were (i) 4-6 weeks after transplanting and (ii) during the 12th weeks of crop growth.

- In maize for example, during the first 2-3 weeks of emergence, weeds often completed 15-18% of their total growth, while maize put up only 2-3% of growth. Such observations have provided a basis in favour of early season weeding to harvest acceptable yields.
- Researcher found that 4 to 16 weeks period after planting sugarcane critical for competition weeds.
- In potato, weeding was found most essential between 2 and 4 weeks after planting. Delayed weeding caused considerable shrinkage in tuber yields.

Table: Crops with an Apparent Critical Period for Weed Competition.

Crop	Weed-free weeks required	Weeks of competition tolerated
Maize	3-5	3-6
Rice	4-6	4-9
Soybean	2-4 after planting	4-8 after planting
Potato	4-6	4-9

Factors Affected Weed Crop Competition

Competition depends on four interrelated factors:

1. Timing of weed emergence: The first plant that effectively obtains water, nutrients and sunlight from a site and becomes established at that site has distinct competitive advantages over plants that develop later. The effect of a weed competition is greatest when the crop is young, since this is the stage which plant growth is inhibited most by inadequate light, water and nutrients. Crop yields are much more reduced by early season weed competition than by later season competition.

2. Growth form: Growth form is manifested in two major parts i.e.

- Growth habit: Extent of root development, height, leaf area, amount of branching
- Growth rate: Those which can develop canopy very rapidly over another, has definite advantage of shading over the second plant communities.

3. Weed Density: The numerical superiority that weeds exhibit greatly reduces the availability of water, nutrients and light to crop plants and accounts for much of what we consider to be weed competition. Increase in crop population density distributes available resources among the crop community, but increase in weed population diverts available resources from the crop communities. For example: 1 kg increase in weed dry matter = 1 kg loss in crop dry matter. Weed density is generally higher in distributed or agricultural soil than in undistributed soils.

4. Duration of weed growth: The duration of weed growth is equally important with all other factors. If weeds are allowed to grow for an extended period crop yield may be drastically reduced. Weeds that are not controlled within 2-3 weeks of emergence usually affect crop yield. This is particularly important for upland rainfed crops i.e. aus rice, jute etc. In most crops weed infestation during the first 3-8 weeks is very critical which is termed as "Critical period" of weed infestation. Crop fields must be kept weed free during this period.

5. Characteristics of Weed species: Weeds differ in their ability to compete with crops at similar density levels. This is primarily because of differences in their growth habits and to some extent in the allelopathic effect they may exert on the germination and growth of neighbouring crop plants. Zimadahl and Fertig (1967) found Brassica spp. (Wild mustard) reduced the sugarbeet yield much more than Setaria glauca (Yelow foxtail). In dry areas perennial weeds like Cirsium arvense (Canada thistle) and Convolvulus arvensis (Bindweed) have been found more competitive than the annual weed species because of their deep roots and early, heavy shoots growth.

6. Characteristics of crop species: Crops and their varieties differ in their competing ability with weeds. Several researchers are available to differentiate crop species and varieties in this respect. Among winter grasses, for example, the decreasing order of weed competing ability is as barley, rye, wheat and oat. High tolerance of barley to competition from weeds is assigned to its ability to develop more extensive roots during its initial three weeks growth period than the other grains.

- Boro Rice: Do not appreciably suffer from weed competition due to standing water throughout the growth period, particularly during the critical periods of weed infestation, i.e. seedling establishment, panicle initiation, flowering stages etc.
- Aman Rice: Do not usually suffer at earlier stages until warming of soil temperature in later growth stages, which causes weeds to germinate or being rapid growth.
- Aus Rice: Most sensitive to weed infestation. Suffers weed competition from very beginning. Cost of production is high due to intensive weed management. Entire crop failure is possible if weed control is not done timely and properly.
- Onion: Slow growing, never forms a canopy, poor competitor to late germinating weeds. Critical period of crop-weed competition was found to be 20-60 DAT. Early infestation of weeds in onion is one of the major constraints limiting the establishment of the crop and thereafter its production.
- Field Pea: The critical period for crop-weed competition was observed to be between 30–60 days after sowing when the crop should be kept free from weeds to prevent the potential yield loss and to economize weeding in fieldpea.

List of Characteristics Associated with Competitive Plants

Shoot Characteristics

- Rapid expansion of tall, foliar canopy.
- Horizontal leaves under overcast conditions and obliquely slanted leaves (plagiotropic) under sunny conditions.
- Large leaves.
- A C4 photosynthetic pathway and low leaf transmissivity of light.
- Leaves forming a mosaic leaf arrangement for best light interception.
- A climbing habit.

- A high allocation of dry matter to build a tall stem.
- Rapid extension in response to shading.

Root Characteristics

- Early and fast root penetration of a large soil area.
- High root density/soil volume.
- High root–shoot ratio.
- High root length per root weight.
- High proportion of actively growing roots.
- Long and abundant root hairs.
- High uptake potential for nutrients and water.

Weed-crop Interactions

The mechanisms of interaction among weeds and crops can be also grouped into several general categories:

Interference

- Competitive ability:
    - Superior growth rates.
    - Internode elongation for enhanced height.
- Aggression and offense:
    - Nutrient luxury consumption: Common lambsquarters.
    - Allelopathy.
    - Parasitism (e.g. dodder; Orabanche).

Coexistence-cohabitation-synergism

- Escape and avoidance: Ecological combining ability and niche diversification:
    - Life history timing (e.g. Recruitment timing, seed dormancy.
    - Spatial patterning.
- Tolerance and defense: Of stress:
    - Crop mimicry.

- Herbicide resistance.
- Shade tolerance.
- Anti-feeding mechanisms: Chemical, structural (e.g. spines).

Competition, Interference and Coexistence

The struggle for existence among weed plants starts at the seedling stage, sometimes earlier, sometimes a little later. Once emerged weed plants grow rapidly and begin to interact with nearby plants (neighbors), either of the same species in the population or with other species (weeds, crops, other plants of the community). As they grow several things happen simultaneously: The compete for limited amounts of resources; their plant number density affects their growth and development; their presence acts as a feedback influence on subsequent seedling recruitment.

Modify the Environment

Interactions dominate most field situations, and the search for unique factors of competiton (beyond death or reduced growth) may not be very sensible. Plants modify the environment of their neighbors by changing conditions, reducing the level of available resources or by adding toxins to the environment. These effects can be demonstrated in artificial populations, but demonstrating them in the field has not been successful.

"Plants may modify the environment of their neighbors by changing conditions, reducing the level of available resources or by adding toxins to the environment. All these effects can be shown quite clearly to operate in experiments with artificial populations, but there are probably no examples of plant interactions in the field in which the mechanism has been clearly and unambiguously demonstrated. Interactions dominate most situations that have been analysed in the field and the search for unique factors of competiton may not be very sensible. Death or a reduced growth rate are often attributable to competive efects, but interpretation of competition in the field may depend on the recognition of much more specific symptoms."

Experimental Characterization of Weed-crop Interactions

Attempts to understand the nature of weed-crop interactions is difficult, if not impossible. Experimentally we attempt to ascribe the success of one weed over another, or over a crop, to a particular morphological feature, a particular pattern of life cycle, or a simple physiological trait.

The chance that a seed will produce a plant is affected in many ways by the presence of another plant, including:

- Resource interference:
  - Light: reducing light intensity; changing light quality.
  - Water: transpiring limited water; changing humidity profile.
  - Nutrients: absorbing limited nutrients; providing limited nitrogen.

    - Gases: reducing CO_2 or O_2 levels, or their ratios in the local atmosphere.
    - Temperature: Altering the temperature of the seed environment.
- Predation:
    - Sheltering or excluding predators (or sheltering predators of predators).
    - Favoring or reducing pathogenic activity.
    - Encouraging defecation or urination in the neighbourhood.
    - Providing rubbing posts or play objects and so encouraging local trampling.
- Direct interference, soil activity:
    - Raising the soil level (accumulation of organic matter, litter).
    - Liberating selective toxins (allelopathy).
    - Changing the soil reaction.

Predators are influencing the germination of seed in this field. Many other factors are contributing to the interaction of these plants including the rocky soil environment, the soil pH, the soybean plants (below, top), the quackgrass plant (below, lower left) and atrazine residues around the hole. Determining the exact mechanism of interaction between plants (seeds) in this spot in the field would be very difficult and complex experimentally. Comparisons of particular factors influencing the interaction of species is very difficult, and will usually not reveal the nature how they affect each other:

- Correlation between individual factors is the best we often do experimentally.
- Differences between plants and species include small and large contributions, the size of those factors is often hard to determine.
- The summation of the contributing factors to a plant-plant interaction is often greater than the individual components: synergistic phenomena emerge as a non-linear consequence of their mutual interaction.

The methodological problems of evaluating the nature of plant-plant interactions is revealed by the problems posed in answering component questions, each themselves involving complex interactions:

- Does the interaction take place above or below the soil surface?
- Does the interaction take place because a plant changes the environment of its neighbors?
- Does the interaction take place because a plant deprives its neighbors of resources?
- Does the interaction take place because a plant produces toxic chemicals (or toxic chemical conditions) that harm or kill its neighbors? Conversely, does it take place because a plant can resist or tolerate a toxic chemical?

References

- Chungnam, parkgiung: kocw.net, Retrieved 19, July 2020
- Weed-dispersal: fabioclass.com, Retrieved 23, April 2020
- Weed-seeds-and-their-longevity-6290: no-tillfarmer.com, Retrieved 12, August 2020
- Crop-weed-competetion-934025: weebly.com, Retrieved 30, January 2020
- Mechanisms, competition, lifehistory: agron.iastate.edu, Retrieved 17, May 2020

Chapter 5

Methods for Weed Management

Weed control is concerned with the attempts to control weeds which compete with the desired flora and fauna. Soil steam sterilization is one method which is used to sterilize the soil with steam and to remove many pests such as weeds, bacteria, fungi, etc. The various methods of weed control like aquatic weed harvester, soil steam sterilization, soil solarization, stale seed bed and bioherbicide, etc. are closely examined in this chapter.

Weed Control

Weed control is the botanical component of pest control, which attempts to stop weeds, especially noxious or injurious weeds, from competing with domesticated plants and livestock. Weed control is important in agriculture. Many strategies have been developed in order to contain these plants. Methods include hand cultivation with hoes, powered cultivation with cultivators, smothering with mulch, lethal wilting with high heat, burning, and chemical attack with herbicides (weed killers).

A plant is often termed a "weed" when it has one or more of the following characteristics:

- Little or no recognized value (as in medicinal, material, nutritional or energy).
- Rapid growth and/or ease of germination.
- Competitive with crops for space, light, water and nutrients.

The definition of a weed is completely context-dependent. To one person, one plant may be a weed, and to another person it may be a desirable plant. In one place, a plant may be viewed as a weed, whereas in another place, the same plant may be desirable.

Introduction

Weeds compete with productive crops or pasture, ultimately converting productive land into unusable scrub. Weeds can be poisonous, distasteful, produce burrs, thorns or otherwise interfere with the use and management of desirable plants by contaminating harvests or interfering with livestock.

Weeds compete with crops for space, nutrients, water and light. Smaller, slower growing seedlings are more susceptible than those that are larger and more vigorous. Onions are one of the most vulnerable, because they are slow to germinate and produce slender, upright stems. By contrast broad beans produce large seedlings and suffer far fewer effects other than during periods of water shortage at the crucial time when the pods are filling out. Transplanted crops raised in sterile soil or potting compost gain a head start over germinating weeds.

Weeds also vary in their competitive abilities and according to conditions and season. Tall-growing vigorous weeds such as fat hen (*Chenopodium album*) can have the most pronounced effects on adjacent crops, although seedlings of fat hen that appear in late summer produce only small plants. Chickweed (*Stellaria media*), a low growing plant, can happily co-exist with a tall crop during the summer, but plants that have overwintered will grow rapidly in early spring and may swamp crops such as onions or spring greens.

The presence of weeds does not necessarily mean that they are damaging a crop, especially during the early growth stages when both weeds and crops can grow without interference. However, as growth proceeds they each begin to require greater amounts of water and nutrients. Estimates suggest that weed and crop can co-exist harmoniously for around three weeks before competition becomes significant. One study found that after competition had started, the final yield of onion bulbs was reduced at almost 4% per day.

Perennial weeds with bulbils, such as lesser celandine and oxalis, or with persistent underground stems such as couch grass (*Agropyron repens*) or creeping buttercup (*Ranunculus repens*) store reserves of food, and are thus able to grow faster and with more vigour than their annual counterparts. Some perennials such as couch grass exude allelopathic chemicals that inhibit the growth of other nearby plants.

Weeds can also host pests and diseases that can spread to cultivated crops. Charlock and Shepherd's purse may carry clubroot, eelworm can be harboured by chickweed, fat hen and shepherd's purse, while the cucumber mosaic virus, which can devastate the cucurbit family, is carried by a range of different weeds including chickweed and groundsel.

Insect pests often do not attack weeds. However pests such as cutworms may first attack weeds then move on to cultivated crops.

Some plants are considered weeds by some farmers and crops by others. Charlock, a common weed in the southeastern US, are weeds according to row crop growers, but are valued by beekeepers, who seek out places where it blooms all winter, thus providing pollen for honeybees and other pollinators. Its bloom resists all but a very hard freeze, and recovers once the freeze ends.

Weed Propagation

Seeds

Annual and biennial weeds such as chickweed, annual meadow grass, shepherd's purse, groundsel, fat hen, cleaver, speedwell and hairy bittercress propagate themselves by seeding. Many produce huge numbers of seed several times a season, some all year round. Groundsel can produce 1000 seed, and can continue right through a mild winter, whilst Scentless Mayweedproduces over 30,000 seeds per plant. Not all of these will germinate at once, but over several seasons, lying dormant in the soil sometimes for years until exposed to light. Poppy seed can survive 80–100 years, dock 50 or more. There can be many thousands of seeds in a square foot or square metre of ground, thus and soil disturbance will produce a flush of fresh weed seedlings.

Subsurface/Surface

The most persistent perennials spread by underground creeping rhizomes that can regrow from a tiny fragment. These include couch grass, bindweed, ground elder, nettles, rosebay willow herb, Japanese knotweed, horsetail and bracken, as well as creeping thistle, whose tap roots can put out lateral roots. Other perennials put out runners that spread along the soil surface. As they creep they set down roots, enabling them to colonise bare ground with great rapidity. These include creeping buttercup and ground ivy. Yet another group of perennials propagate by stolons- stems that arch back into the ground to reroot. The most familiar of these is the bramble.

Methods

Weed control plans typically consist of many methods which are divided into biological, chemical, cultural, and physical/mechanical control.

Pesticide-free thermic weed control with a weed burner on a potato field in Dithmarschen.

Physical/Mechanical Methods

Coverings

In domestic gardens, methods of weed control include covering an area of ground with a material that creates a hostile environment for weed growth, known as a *weed mat*.

Several layers of wet newspaper prevent light from reaching plants beneath, which kills them. Daily saturating the newspaper with water plant decomposition. After several weeks, all germinating weed seeds are dead.

In the case of black plastic, the greenhouse effect kills the plants. Although the black plastic sheet is effective at preventing weeds that it covers, it is difficult to achieve complete coverage. Eradicating persistent perennials may require the sheets to be left in place for at least two seasons.

Some plants are said to produce root exudates that suppress herbaceous weeds. *Tagetes minuta* is claimed to be effective against couch and ground elder, whilst a border of comfrey is also said to act as a barrier against the invasion of some weeds including couch. A 5–10 centimetres (2.0–3.9 in)} layer of wood chip mulch prevents most weeds from sprouting.

Gravel can serve as an inorganic mulch.

Irrigation is sometimes used as a weed control measure such as in the case of paddy fields to kill any plant other than the water-tolerant rice crop.

Manual Removal

Weeds are removed manually in large parts of India.

Many gardeners still remove weeds by manually pulling them out of the ground, making sure to include the roots that would otherwise allow them to resprout.

Hoeing off weed leaves and stems as soon as they appear can eventually weaken and kill perennials, although this will require persistence in the case of plants such as bindweed. Nettle infestations can be tackled by cutting back at least three times a year, repeated over a three-year period. Bramble can be dealt with in a similar way.

Tillage

Ploughing includes tilling of soil, intercultural ploughing and summer ploughing. Ploughing uproots weeds, causing them to die. In summer ploughing is done during deep summers. Summer ploughing also helps in killing pests.

Mechanical tilling can remove weeds around crop plants at various points in the growing process.

Thermal

Several thermal methods can control weeds.

Flame weeders use a flame several centimeters away from the weeds to give them a sudden and severe heating. The goal of flame weeding is not necessarily burning the plant, but rather causing a lethal wilting by denaturing proteins in the weed. Similarly, hot air weeders can heat up the seeds to the point of destroying them. Flame weeders can be combined with techniques such as stale seedbeds (preparing and watering the seedbed early, then killing the nascent crop of weeds that springs up from it, then sowing the crop seeds) and preemergence flaming (doing a flame pass against weed seedlings after the sowing of the crop seeds but before those seedlings emerge from the soil—a span of time that can be days or weeks).

Hot foam (foamstream) causes the cell walls to rupture, killing the plant. Weed burners heat up soil quickly and destroy superficial parts of the plants. Weed seeds are often heat resistant and even react with an increase of growth on dry heat.

Since the 19th century soil steam sterilization has been used to clean weeds completely from soil. Several research results confirm the high effectiveness of humid heat against weeds and its seeds.

Soil solarization in some circumstances is very effective at eliminating weeds while maintaining grass. Planted grass tends to have a higher heat/humidity tolerance than unwanted weeds.

Seed Targeting

In 1998, the Australian Herbicide Resistance Initiative (AHRI), debuted. gathered fifteen scientists and technical staff members to conduct field surveys, collect seeds, test for resistance and study the biochemical and genetic mechanisms of resistance. A collaboration with DuPont led to a mandatory herbicide labeling program, in which each mode of action is clearly identified by a letter of the alphabet.

The key innovation of the AHRI approach has been to focus on weed seeds. Ryegrass seeds last only a few years in soil, so if farmers can prevent new seeds from arriving, the number of sprouts will shrink each year. Until the new approach farmers were unintentionally helping the seeds. Their combines loosen ryegrass seeds from their stalks and spread them over the fields. In the mid-1980s, a few farmers hitched covered trailers, called "chaff carts", behind their combines to catch the chaff and weed seeds. The collected material is then burned.

An alternative is to concentrate the seeds into a half-meter-wide strip called a windrow and burn the windrows after the harvest, destroying the seeds. Since 2003, windrow burning has been adopted by about 70% of farmers in Western Australia.

Yet another approach is the Harrington Seed Destructor, which is an adaptation of a coal pulverizing cage mill that uses steel bars whirling at up to 1500 rpm. It keeps all the organic material in the field and does not involve combustion, but kills 95% of seeds.

Cultural Methods

Stale Seed Bed

Another manual technique is the 'stale seed bed', which involves cultivating the soil, then leaving it fallow for a week or so. When the initial weeds sprout, the grower lightly hoes them away before planting the desired crop. However, even a freshly cleared bed is susceptible to airborne seed from elsewhere, as well as seed carried by passing animals on their fur, or from imported manure.

Buried Drip Irrigation

Buried drip irrigation involves burying drip tape in the subsurface near the planting bed, thereby limiting weeds access to water while also allowing crops to obtain moisture. It is most effective during dry periods.

Crop Rotation

Rotating crops with ones that kill weeds by choking them out, such as hemp, *Mucuna pruriens*, and other crops, can be a very effective method of weed control. It is a way to avoid the use of herbicides, and to gain the benefits of crop rotation.

Biological Methods

A biological weed control regiment can consist of biological control agents, bioherbicides, use of grazing animals, and protection of natural predators.

Animal Grazing

Companies using goats to control and eradicate leafy spurge, knapweed, and other toxic weeds have sprouted across the American West.

Chemical Methods

"Organic" Approaches

Weed control, circa 1930-40s.

A mechanical weed control device: the diagonal weeder.

Organic weed control involves anything other than applying manufactured chemicals. Typically a combination of methods are used to achieve satisfactory control.

Sulfur in some circumstances is accepted within British Soil Association standards.

Herbicides

The above described methods of weed control use no or very limited chemical inputs. They are preferred by organic gardeners or organic farmers.

However weed control can also be achieved by the use of herbicides. Selective herbicides kill certain targets while leaving the desired crop relatively unharmed. Some of these act by interfering with the growth of the weed and are often based on plant hormones. Herbicides are generally classified as follows:

- Contact herbicides destroy only plant tissue that contacts the herbicide. Generally, these are the fastest-acting herbicides. They are ineffective on perennial plants that can re-grow from roots or tubers.
- Systemic herbicides are foliar-applied and move through the plant where they destroy a greater amount of tissue. Glyphosate is currently the most used systemic herbicide.
- Soil-borne herbicides are applied to the soil and are taken up by the roots of the target plant.
- Pre-emergent herbicides are applied to the soil and prevent germination or early growth of weed seeds.

In agriculture large scale and systematic procedures are usually required, often by machines, such as large liquid herbicide 'floater' sprayers, or aerial application.

Bradley Method

Which uses ecological processes to do much of the work. Perennial weeds also propagate by seeding; the airborne seed of the dandelion and the rose-bay willow herb parachute far and wide. Dandelion and dock also put down deep tap roots, which, although they do not spread underground, are able to regrow from any remaining piece left in the ground.

Hybrid

One method of maintaining the effectiveness of individual strategies is to combine them with others that work in complete different ways. Thus seed targeting has been combined with herbicides. In Australia seed management has been effectively combined with trifluralin and clethodim.

Resistance

Resistance occurs when a target adapts to circumvent a particular control strategy. It affects not only weed control,but antibiotics, insect control and other domains. In agriculture is mostly considered in reference to pesticides, but can defeat other strategies, e.g., when a target species becomes more drought tolerant via selection pressure.

Farming Practices

Herbicide resistance recently became a critical problem as many Australian sheep farmers switched to exclusively growing wheat in their pastures in the 1970s. In wheat fields, introduced varieties of ryegrass, while good for grazing sheep, are intense competitors with wheat. Ryegrasses produce so many seeds that, if left unchecked, they can completely choke a field. Herbicides provided excellent control, while reducing soil disrupting because of less need to plough. Within little more than a decade, ryegrass and other weeds began to develop resistance. Australian farmers evolved again and began diversifying their techniques.

In 1983, patches of ryegrass had become immune to Hoegrass, a family of herbicides that inhibit an enzyme called acetyl coenzyme A carboxylase.

Ryegrass populations were large, and had substantial genetic diversity, because farmers had planted many varieties. Ryegrass is cross-pollinated by wind, so genes shuffle frequently. Farmers sprayed inexpensive Hoegrass year after year, creating selection pressure, but were diluting the herbicide in order to save money, increasing plants survival. Hoegrass was mostly replaced by a group of herbicides that block acetolactate synthase, again helped by poor application practices. Ryegrass evolved a kind of "cross-resistance" that allowed it to rapidly break down a variety of herbicides. Australian farmers lost four classes of herbicides in only a few years. As of 2013 only two herbicide classes, called Photosystem II and long-chain fatty acid inhibitors, had become the last hope.

Aquatic Weed Harvester

An aquatic weed harvester in action on a freshwater lake.

An aquatic weed harvester.

An aquatic weed harvester, also known as a water mower, mowing boat and weed cutting boat, is an aquatic machine specifically designed for inland watercourse management to cut and harvest underwater weeds, reeds and other aquatic plant life. The action of removing aquatic plant life in such a manner has been referred to as "aquatic harvesting".

Overview

Water is an important resource and in many countries, waterways are increasingly clogged by aquatic plant growth. This is particularly so in tropical countries where warmer water means the plants grow more quickly, and increasing run-off of fertilisers and effluent has exacerbated the problem. Irrigation ditches and pumps can become overgrown with vegetation, power station and factory water intakes can get blocked, boats can get hindered, fish stocks can be disrupted, and water moves more slowly, resulting in greater evapotranspiration and a greater risk of flooding. In some large irrigation projects in India, canals have become so overgrown with vegetation that water flow has been reduced to a fifth of its previous amount. In Bangladesh, floodwater has washed mats of water hyacinth onto paddy fields, overwhelming the emerging rice crops. Small fish can become entangled in excessive algal growth.

A pond in Odessa, Ukraine covered with water lettuce (*Pistia stratiotes*).

Rice is the main aquatic plant grown for human food, but smaller areas of watercress and water chestnut are also cultivated. In their native environments, aquatic weeds are part of a balanced ecosystem, and it is mainly introduced species of water plant that become invasive and cause problems by congesting water bodies. The worst culprits, found in both temperate and tropical waterways, are floating plants such as water hyacinth, water lettuce and *Salvinia*, fully submerged rooting plants such as *Hydrilla* and water milfoil and rooting plants that reach the surface such as cattail, papyrus, bulrush and reed.

Weed Harvesting Equipment

Weed cutting boats are developed to enable the maintenance of canals, lakes and rivers and to remove excessive aquatic life such as algae and other plants that may negatively affect a waterway's ecology. Mechanical harvesters are large floating machines that have underwater cutting blades that sever the stems of underwater plants, gather the weeds and raise them on conveyor belts, storing the vegetation on board in a hold. Periodically this is discharged to a barge or an onshore facility. The harvested product can be composted, sent to a landfill site or used in land reclamation.

In developing countries aquatic vegetation may be harvested by hand or by net from the shore, cut and harvested by boat and lifted ashore by hand, crane, pump or conveyor system. The harvested vegetation may be used for the feeding of livestock. To reduce the high moisture content and to make it easier to transport, the weed can be chopped and pressed. Other uses to which the harvested vegetation can be put include ensiling the material for livestock fodder, adding it to the soil as a bulky organic fertilizer, manufacturing the raw material into pulp, paper or fibre, and fermenting it to produce methane for energy production.

Advantages and Disadvantages

Mechanical harvesters can be effective at clearing aquatic weeds but the machines are expensive and the process may need to be repeated several times in a growing season. Small fragments of weed remain in the water and may spread to other locations thereby aiding in the dispersal of invasive species. Some areas may be too shallow for the mechanical harvester and it may be unable to access restricted locations. Submerged tree stumps can damage the machine. An alternative to mechanical harvesting is the use of herbicides, which are easy to apply and less expensive, but may have unwanted impacts on the environment.

By Region

Africa

Water hyacinth growing in Lake Victoria at Kisumu.

In June 2015 in Uganda, use of the aquatic weed harvester was recommended by the country's Ministry of Agriculture to reduce water hyacinth growth in Lake Victoria, which has caused a scarcity of fish in the lake. The scarcity of fish has negatively-affected the livelihood of locals who live in the lake's region. Additional types of machines were recommended to address the problem, which were the hydraulic harvester, take out elevator and dredger.

India

In August 2015 in Hyderabad, India, an aquatic weed harvester was used to remove algal weeds and trash from Hussainsagar Lake. An amphibious hydraulic excavator was also used. The work was performed to address pollution problems at the lake, which had accumulated significant algae and rubbish such as plastic bags and plastic waste, food wrappers, and various garbage, some of which was floating atop the lake.

A mowing boat.

United States

Some U.S. companies manufacture aquatic weed harvesters. One such machine manufactured by a U.S. company can cut around one-half an acre of weeds a day, and costs over $100,000.

Tampa, Florida used an aquatic weed harvester in 2013 to clear aquatic plant life from lakes in the area.

In 2012 in Leoni, Michigan, an aquatic weed harvester was used to remove algae from Center Lake. The algae appeared to have roots, and other techniques such as attempting to kill it with chemicals were unsuccessful. The machine used collects algae and weeds that live at the bottom of the lake, removing them from the waterway.

Soil Steam Sterilization

Soil steam sterilization (soil steaming) is a farming technique that sterilizes soil with steam in open fields or greenhouses. Pests of plant cultures such as weeds, bacteria, fungi and viruses are killed through induced hot steam which causes their cell structure to physically degenerate. Biologically, the method is considered a partial disinfection. Important heat-resistant, spore-forming bacteria survive and revitalize the soil after cooling down. Soil fatigue can be cured through the release of nutritive substances blocked within the soil. Steaming leads to a better starting position, quicker growth and strengthened resistance against plant disease and pests. Today, the application of hot steam is considered the best and most effective way to disinfect sick soil, potting soil and compost. It is being used as an alternative to bromomethane, whose production and use was curtailed by the Montreal Protocol. "Steam effectively kills pathogens by heating the soil to levels that cause protein coagulation or enzyme inactivation."

Benefits of Soil Steaming

Soil sterilization provides secure and quick relief of soils from substances and organisms harmful to plants such as:

- Bacteria.

- Viruses.
- Fungi.
- Nematodes.
- Other Pests.

Further positive effects are:

- All weed and weed seeds are killed.
- Significant increase of crop yields.
- Relief from soil fatigue through activation of chemical – biological reactions.
- Blocked nutritive substances in the soil are tapped and made available for plants.
- Alternative to Methyl Bromide and other critical chemicals in agriculture.

Steaming With Superheated Steam

Through modern steaming methods with superheated steam at 180–200 °C, an optimal soil disinfection can be achieved. Soil only absorbs a small amount of humidity. Micro organisms become active once the soil has cooled down. This creates an optimal environment for instant tillage with seedlings and seeds. Additionally the method of integrated steaming can promote a target-oriented resettlement of steamed soil with beneficial organisms. In the process, the soil is first freed from all organisms and then revitalized and microbiologically buffered through the injection of a soil activator based on compost which contains a natural mixture of favorable microorganisms (e.g. Bacillus subtilis, etc.).

Different types of such steam application are also available in practice, including substrate steaming and surface steaming.

Surface Steaming

Several methods for surface steaming are in use amongst which are: area sheet steaming, the steaming hood, the steaming harrow, the steaming plough and vacuum steaming with drainage pipes or mobile pipe systems.

In order to pick the most suitable steaming method, certain factors have to be considered such as soil structure, plant culture and area performance. At present, more advanced methods are being developed, such as sandwich steaming or partially integrated sandwich steaming in order to minimize energy and cost as much as possible.

Sheet Steaming

Large area sheet steaming in greenhouses using a steam injector.

Surface steaming with special sheets (sheet steaming) is a method which has been established for decades in order to steam large areas reaching from 15 to 400 m^2 in one step. If properly applied, sheet steaming is simple and highly economic. The usage of heat resistant, non-decomposing insulation fleece saves up to 50% energy, reduces the steaming time significantly and improves penetration. Single working step areas up to 400 m² can be steamed in 4–5 hours down to 25–30 cm depth / 90°C. The usage of heat resistant and non-decomposing synthetic insulation fleece, 5 mm thick, 500 gr / m², can reduce steaming time by about 30%. Through a steam injector or a perforated pipe, steam is injected underneath the sheet after it has been laid out and weighted with sand sacks.

The area performance in one working step depends on the capacity of the steam generator (e.g. steam boiler):

Steam capacity kg/h:	100	250	300	400	550	800	1000	1350	2000
Area m²:	15-20	30-50	50-65	60-90	80-120	130-180	180-220	220-270	300-400

The steaming time depends on soil structure as well as outside temperature and amounts to 1-1.5 hours per 10 cm steaming depth. Hereby the soil reaches a temperature of about 85°C. Milling for soil loosening is not recommended since soil structure may become too fine which reduces its penetrability for steam. The usage of spading machines is ideal for soil loosening. The best results can be achieved if the soil is cloddy at greater depth and granulated at lesser depth.

In practice, working with at least two sheets simultaneously has proven to be highly effective. While one sheet is used for steaming the other one is prepared for steam injection, therefore unnecessary steaming recesses are avoided.

Depth Steaming With Vacuum

Steaming with vacuum which is induced through a mobile or fixed installed pipe system in the depth of the area to be steamed, is the method that reaches the best penetration. Despite high capital cost, the fixed installation of drainage systems is reasonable for intensively used areas since steaming depths of up to 80 cm can be achieved.

In contrast to fixed installed drainage systems, pipes in mobile suction systems are on the surface. A central suction pipeline consisting of zinc-coated, fast-coupling pipes are connected in a regular spacing of 1.50 m and the ends of the hoses are pushed into the soil to the desired depth with a special tool.

The steaming area is covered with a special steaming sheet and weighted all around as with sheet steaming. The steam is injected underneath the sheet through an injector and protection tunnel. While with short areas up to 30 m length steam is frontally injected, with longer areas steam is induced in the middle of the beet using a T-connection branching out to both sides. As soon as the sheet is inflated to approximately 1m by the steam pressure, the suction turbine is switched on. First, the air in the soil is removed via the suction hoses. A vacuum is formed and the steam is pulled downward.

During the final phase, when the required steaming depth has been reached, the ventilator runs non-stop and surplus steam is blown out. To ensure that this surplus steam is not lost, it is fed back under the sheet.

As with all other steaming systems, a post-steaming period of approximately 20–30 minutes is required. Steaming time is approximately 1 hour per 10 cm steaming depth. The steam requirement is approximately 7–8 kg/m^2.

The most important requirement, as with all steaming systems, is that the soil is well loosened before steaming, to ensure optimal penetration.

Negative Pressure Technique

"Negative Pressure technique generates appropriate soil temperature at a 60 cm depth and complete control of nematodes, fungi and weeds is achieved. In this technique, the steam is introduced under the steaming sheath and forced to enter the soil profile by a negative pressure. The negative pressure is created by a fan that sucks the air out of the soil through buried perforated polypropylene pipes. This system requires a permanent installation of perforated pipes into the soil, at a depth of at least 60 cm to be protected from plough."

Steaming with Hoods

Half automatic steaming hood with three wings in greenhouse.

A steaming hood is a mobile device consisting of corrosion-resistant materials such as aluminum, which is put down onto the area to be steamed. In contrast to sheet steaming, cost-intensive working steps such as laying out and weighting the sheets don't occur, however the area steamed per working step is smaller in accordance to the size of the hood.

Outdoors, a hood is positioned either manually or via tractor with a special pre-stressed 4 point suspension arm. Steaming time amounts to 30 min for a penetration down to 25 cm depth. Hereby a temperature of 90°C can be reached. In large stable glasshouses, the hoods are attached to tracks. They are lifted and moved by pneumatic cylinders. Small and medium-sized hoods up to 12m^2 are lifted manually using a tipping lever or moved electrically with special winches.

Combined Surface and Depth Injection of Steam (Sandwich Steaming)

Sandwich steaming, which was developed in a project among DEIAFA, University of Turin (Italy, www.deiafa.unito.it) and Ferrari Costruzioni Meccaniche, represents a combination of depth and surface steaming, offers an efficient method to induce hot steam into the soil. The steam is simultaneously pushed into the soil from the surface and from the depth. For this purpose, the area, which must be equipped with a deep steaming injection system, is covered with a steaming hood. The steam enters the soil from the top and the bottom at the same time. Sheets are not suitable, since a high pressure up to 30 mm water column arises underneath the cover.

Sandwich steaming offers several advantages. On the one hand, application of energy can be increased to up to 120 kg steam per m^2/h. In comparison to other steaming methods up to 30% energy savings can be achieved and the usage of fuel (e.g. heating oil) accordingly decreases. The increased application of energy leads to a quick heating of the soil which reduces the loss of heat. On the other hand, only half of the regular steaming time is needed.

Sandwich steaming machine model Sterilter constructed by Ferrari Costruzioni Meccaniche equipped with MSD/ moeschle steam boiler.

Comparison of sandwich steaming with other steam injection methods relating to steam output and energy demand(*):

Steaming method	max. steam output	energy demand (*)
Sheet steaming	6 kg/m^2h	about 100 kg $steam/m^3$
Depth steaming (Sheet + vacuum)	14 kg/m^2h	about 120 kg $steam/m^3$
Hood steaming (Alu)	30 kg/m^2h	about 80 kg $steam/m^3$
Hood steaming (Steel)	50 kg/m^2h	about 75 kg $steam/m^3$
Sandwich steaming	120 kg/m^2h	about 60 kg $steam/m^3$

(*) in soil max 30% moisture

Clearly, Sandwich steaming reaches the highest steam output at the lowest energy demand.

Partially Integrated Sandwich Steaming

The partial integrated Sandwich steaming is an advanced combined method for steaming merely the areas which shall be planted and purposely leaving out those areas which shall not be used. In order to avoid risk of re-infection of steamed areas with pest from unsteamed areas, beneficial organisms can directly be injected into the hygenized soil via a soil activator (e.g. special compost). The partial sandwich steaming unlocks further potential savings in the steaming process.

Container / Stack Steaming

Stack steaming is used when thermically treating compost and substrates such as turf. Depending on the amount, the material to be steamed is piled up to 70 cm height in steaming boxes or in small dump trailers. Steam is evenly injected via manifolds. For huge amounts, steaming containers and soil boxes are used which are equipped with suction systems to improve steaming results. Midget amounts can be steamed in special small steaming devices.

The amount of soil steamed should be tuned in a way that steaming time amounts to at most 1.5 h in order to avoid large quantities of condensed water in the bottom layers of the soil.

Steam Output kg/h:	100	250	300	400	550	800	1000	1350	2000
m^3/h about:	1.0-1.5	2.5-3.0	3.0-3.5	4.0-5.0	5.5-7.0	8.0-10.0	10.0-13.0	14.0-18.0	20.0-25.0

In light substrates, such as turf, the performance per hour is significantly higher.

History

Ancient civilizations in India and Egypt used steam, generated through the targeted usage of incident solar radiation on watered top soil, to sanitize and revive their arable land.

Modern soil steam sterilization was first discovered in 1888 (by Frank in Germany) and was first commercially used in the United States (by Rudd) in 1893 (Baker 1962). Since then, a wide variety of steam machines have been built to disinfest both commercial greenhouse and nursery field soils (Grossman and Liebman 1995). In the 1950s, for example, steam sterilization technologies expanded from disinfestation of potting soil and greenhouse mixes to commercial production of steam rakes and tractor-drawn steam blades for fumigating small acres of cut flowers and other high-value field crops (Langedijk 1959). Today, even more effective steam technologies are being developed.

Application of Hot Steam

- In horticulture as well as nurseries for sterilization of substrates and top soil.
- In agriculture for sterilization and treatment of food waste for pig fattening and heating of molasses.
- In mushroom cultivation for pasteurization of growing rooms, sterilization of top soil and combined application as heating.
- In wineries as combination boiler for sterilization and cleaning of storage tanks, tempering of mash and for warm water generation.

Soil Solarization

Soil solarization is an environmentally friendly method of using solar power for controlling pests such as soilborne plant pathogens including fungi, bacteria, nematodes, and insect and mite pests along with weed seed and seedlings in the soil by mulching the soil and covering it with tarp, usually with a transparent polyethylene cover, to trap solar energy. It may also describe methods of decontaminating soil using sunlight or solar power. This energy causes physical, chemical, and biological changes in the soil.

Soil Disinfestation

Soil solarization is a relatively new soil disinfestation method, first described in extensive scientific detail by Katan et al. in 1976, presenting the results of a series of studies performed under field conditions, initiated in 1973, for controlling soilborne pathogens and weeds, mostly as a pre-planting soil treatment. Soil is mulched and then covered with transparent polyethylene during the hot season, thereby heating it and killing the pests.

Soil Decontamination

A 2008 study used a solar cell to generate an electric field for electrokinetic (EK) remediation of cadmium-contaminated soil. The solar cell could drive the electromigration of cadmium in contaminated soil, and the removal efficiency that was achieved by the solar cell was comparable with that achieved by conventional power supply.

In Korea, various remediation methods of soil slurry and groundwater contaminated with benzene at a polluted gas station site were evaluated, including a solar-driven, photocatalyzed reactor system along with various advanced oxidation processes (AOP). The most synergistic remediation method incorporated a solar light process with TiO2 slurry and H2O2 system, achieving 98% benzene degradation, a substantial increase in the removal of benzene.

History

Attempts were made to use solar energy for controlling disease agents in soil and in plant material already in the ancient civilization of India. In 1939, Groashevoy, who used the term "solar energy for sand disinfection," controlled Thielaviopsis basicola upon heating the sand by exposure to direct sunlight.

Soil solarization is the third approach for soil disinfestation; the two other main approaches, soil steaming and fumigation; were developed at the end of the 19th century. The idea of solarization was based on observations by extension workers and farmers in the hot Jordan Valley, who noticed the intensive heating of the polyethylene-mulched soil. The involvement of biological control mechanisms in pathogen control and the possible implications were indicated in the first publication, noticing the very long effect of the treatment. In 1977, American scientists from the University of California at Davis reported the control of Verticillium in a cotton field, based on studies started in 1976, thus denoting, for the first time, the possible wide applicability of this method.

The use of polyethylene for soil solarization differs in principle from its traditional agricultural use. With solarization, soil is mulched during the hottest months (rather than the coldest, as in conventional plasticulture which is aimed at protecting the crop) in order to increase the maximal temperatures in an attempt to achieve lethal heat levels.

In the first 10 years following the influential 1976 publication, soil solarization was investigated in at least 24 countries and has been now been applied in more than 50, mostly in the hot regions, although there were some important exceptions. Studies have demonstrated effectiveness of solarization with various crops, including vegetables, field crops, ornamentals and fruit trees, against many pathogens, weeds and a soil arthropod. Those pathogens and weeds which are not controlled by solarization were also detected. The biological, chemical and physical changes that take in solarized soil during and after the solarization have been investigated, as well as the interaction of solarization with other methods of control. Long-term effects including biological control and increased growth response were verified in various climatic regions and soils, demonstrating the general applicability of solarization. Computerized simulation models have been developed to guide researchers and growers whether the ambient conditions of their locality are suitable for solarization.

Studies of the improvement of solarization by integrating it with other methods or by solarizing in closed glasshouses, or studies concerning commercial application by developing mulching machines were also carried out.

The use of solarization in existing orchards (e.g. controlling Verticillium in pistachio plantations) is an important deviation from the standard preplanting method and was reported as early as 1979.

Stale Seed Bed

A false or stale seed bed is a useful weed control technique which involves creating a seedbed some weeks before seed is due to be sown. Preparation of such seedbed makes sure that any weed seeds that have been disturbed and brought to the soil surface during cultivation will thus have a chance to germinate, and can then be hoed off or eliminated with the use of a flame weeder before sowing of the actual crop is carried out.

The technique can be utilized in early spring, when the weather is still too cold for proper seed germination. Several passes are made with a rototiller or plow, then weed seeds are allowed to germinate as weather permits.

By tilling, the farmer increases the chance of weed seed germination by the same method as one would for favorable vegetable/crops: the fine soil allows weed seed to grow rapidly by allowing the seed to open and the roots to spread easier than in compacted soil. Deep tilling will also bring dormant seed to the surface for germination. Some species of plant are known for seeds that can lay deeply buried in the soil for years before favorable conditions allow germination.

Timing is important, however. Weed seeds must be tilled/howed or otherwise destroyed before they themselves can create new seeds. By destroying them early, the farmer eliminates most of that season's annual weeds. Turning the dead weeds back into the soil also increases soil nutrient content, although this difference is slight.

In many cases, several tillings are done, perhaps every two weeks beginning in very early spring. This allows more and more weed seeds to germinate only to be killed off later. This eliminates more weeds, but care must be used to not delay planting of a desirable crop later than the crop needs for a successful season's growth.

After several years, most, if not all, weeds can be eliminated from the seed bank in the soil. In some cases the effect can be noticed in the same year the process is first carried out.

If the weed patch is vigorous enough, the farmer may mow the patch first before tilling. This allows for easier/quicker decomposition in the soil when the plants are turned under.

Some farmers are noted for applying a light and inexpensive fertilizer mix to the soil to cause even more weed seeds to germinate. Although the merits of this may seem a trifle at first, it can pay dividends in the long run by eliminating seeds earlier that otherwise would have sprouted in later years. This too is open to debate.

Drip Irrigation

An Emitter or dripper in action.

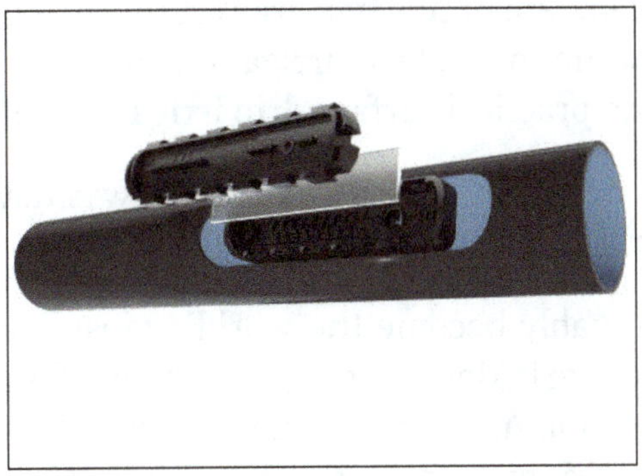

Open pressure compensated dripper.

Drip irrigation is a form of irrigation that saves water and fertilizer by allowing water to drip slowly to the roots of many different plants, either onto the soil surface or directly onto the root zone, through a network of valves, pipes, tubing, and emitters. It is done through narrow tubes that deliver water directly to the base of the plant. It is chosen instead of surface irrigation for various reasons, often including concern about minimizing evaporation.

History

Primitive drip irrigation has been used since ancient times. Fan Sheng-Chih Shu, written in China during the first century BCE, describes the use of buried, unglazed clay pots filled with water as a means of irrigation. Modern drip irrigation began its development in Germany in 1860 when researchers began experimenting with subsurface irrigation using clay pipe to create combination irrigation and drainage systems. Research was later expanded in the 1920s to include the application of perforated pipe systems. The usage of plastic to hold and distribute water in drip irrigation was later developed in Australia by Hannis Thill.

Drip irrigation in Mexico vineyard, 2000.

Usage of a plastic emitter in drip irrigation was developed in Israel by Simcha Blass and his son Yeshayahu. Instead of releasing water through tiny holes easily blocked by tiny particles, water was released through larger and longer passageways by using velocity to slow water inside a plastic emitter. The first experimental system of this type was established in 1959 by Blass who partnered later (1964) with Kibbutz Hatzerim to create an irrigation company called Netafim. Together they developed and patented the first practical surface drip irrigation emitter.

In the United States, the first drip tape, called *Dew Hose*, was developed by Richard Chapin of Chapin Watermatics in the early 1960s.

Modern drip irrigation has arguably become the world's most valued innovation in agriculture since the invention of the impact sprinkler in the 1930s, which offered the first practical alternative to surface irrigation. Drip irrigation may also use devices called micro-spray heads, which spray water in a small area, instead of dripping emitters. These are generally used on tree and vine crops with wider root zones. Subsurface drip irrigation (SDI) uses permanently or temporarily buried dripperline or drip tape located at or below the plant roots. It is becoming popular for row crop irrigation, especially in areas where water supplies are limited or recycled water is used for irrigation. Careful study of all the relevant factors like land topography, soil, water, crop and agro-climatic conditions are needed to determine the most suitable drip irrigation system and components to be used in a specific installation.

Components and Operation

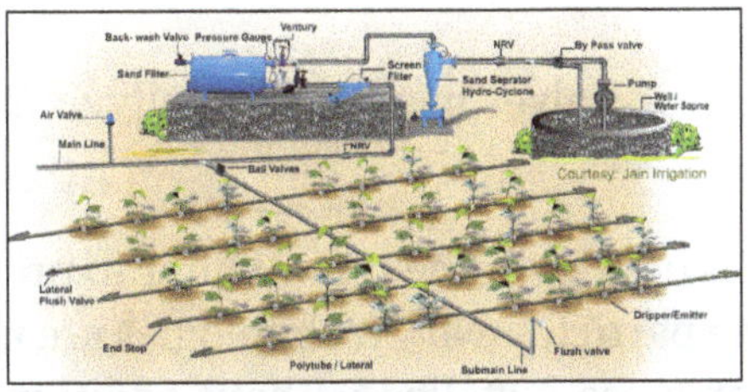

Drip irrigation system layout and its parts.

Water distribution in subsurface drip irrigation.

Nursery flowers watered with drip irrigation in Israel.

Components used in drip irrigation (listed in order from water source) include:

- Pump or pressurized water source.
- Water filter(s) or filtration systems: sand separator, Fertigation systems (Venturi injector) and chemigation equipment (optional).
- Backwash controller (Backflow prevention device).
- Pressure Control Valve (pressure regulator).
- Distribution lines (main larger diameter pipe, maybe secondary smaller, pipe fittings).
- Hand-operated, electronic, or hydraulic control valves and safety valves.
- Smaller diameter polytube (often referred to as “laterals”).
- Poly fittings and accessories (to make connections).

- Emitting devices at plants (emitter or dripper, micro spray head, inline dripper or inline driptube).

In drip irrigation systems, pump and valves may be manually or automatically operated by a controller.

Most large drip irrigation systems employ some type of filter to prevent clogging of the small emitter flow path by small waterborne particles. New technologies are now being offered that minimize clogging. Some residential systems are installed without additional filters since potable water is already filtered at the water treatment plant. Virtually all drip irrigation equipment manufacturers recommend that filters be employed and generally will not honor warranties unless this is done. Last line filters just before the final delivery pipe are strongly recommended in addition to any other filtration system due to fine particle settlement and accidental insertion of particles in the intermediate lines.

Drip and subsurface drip irrigation is used almost exclusively when using recycled municipal waste water. Regulations typically do not permit spraying water through the air that has not been fully treated to potable water standards.

Because of the way the water is applied in a drip system, traditional surface applications of timed-release fertilizer are sometimes ineffective, so drip systems often mix liquid fertilizer with the irrigation water. This is called fertigation; fertigation and chemigation (application of pesticides and other chemicals to periodically clean out the system, such as chlorine or sulfuric acid) use chemical injectors such as diaphragm pumps, piston pumps, or aspirators. The chemicals may be added constantly whenever the system is irrigating or at intervals. Fertilizer savings of up to 95% are being reported from recent university field tests using drip fertigation and slow water delivery as compared to timed-release and irrigation by micro spray heads.

Properly designed, installed, and managed, drip irrigation may help achieve water conservation by reducing evaporation and deep drainage when compared to other types of irrigation such as flood or overhead sprinklers since water can be more precisely applied to the plant roots. In addition, drip can eliminate many diseases that are spread through water contact with the foliage. Finally, in regions where water supplies are severely limited, there may be no actual water savings, but rather simply an increase in production while using the same amount of water as before. In very arid regions or on sandy soils, the preferred method is to apply the irrigation water as slowly as possible.

Pulsed irrigation is sometimes used to decrease the amount of water delivered to the plant at any one time, thus reducing runoff or deep percolation. Pulsed systems are typically expensive and require extensive maintenance. Therefore, the latest efforts by emitter manufacturers are focused toward developing new technologies that deliver irrigation water at ultra-low flow rates, i.e. less than 1.0 liter per hour. Slow and even delivery further improves water use efficiency without incurring the expense and complexity of pulsed delivery equipment.

An emitting pipe is a type of drip irrigation tubing with emitters pre-installed at the factory with specific distance and flow per hour as per crop distance.

An emitter restricts water flow passage through it, thus creating head loss required (to the extent of atmospheric pressure) in order to emit water in the form of droplets. This head loss is achieved by friction / turbulence within the emitter.

Advantages and Disadvantages

Drip irrigation and spare drip irrigation tubes in banana farm at Chinawal, India.

Pot irrigation by On-line drippers.

Pressure compensated integral dripper on soilless without growing channels.

The advantages of drip irrigation are:

- Fertilizer and nutrient loss is minimized due to localized application and reduced leaching.
- Water application efficiency is high if managed correctly.
- Field levelling is not necessary.

- Fields with irregular shapes are easily accommodated.
- Recycled non-potable water can be safely used.
- Moisture within the root zone can be maintained at field capacity.
- Soil type plays less important role in frequency of irrigation.
- Soil erosion is lessened.
- Weed growth is lessened.
- Water distribution is highly uniform, controlled by output of each nozzle.
- Labour cost is less than other irrigation methods.
- Variation in supply can be regulated by regulating the valves and drippers.
- Fertigation can easily be included with minimal waste of fertilizers.
- Foliage remains dry, reducing the risk of disease.
- Usually operated at lower pressure than other types of pressurised irrigation, reducing energy costs.

The disadvantages of drip irrigation are:

M

- Initial cost can be more than overhead systems.
- The sun can affect the tubes used for drip irrigation, shortening their usable life.
- If the water is not properly filtered and the equipment not properly maintained, it can result in clogging.
- For subsurface drip the irrigator cannot see the water that is applied. This may lead to the farmer either applying too much water (low efficiency) or an insufficient amount of water, this is particularly common for those with less experience with drip irrigation.
- Drip irrigation might be unsatisfactory if herbicides or top dressed fertilizers need sprinkler irrigation for activation.
- Drip tape causes extra cleanup costs after harvest. Users need to plan for drip tape winding, disposal, recycling or reuse.
- Waste of water, time and harvest, if not installed properly. These systems require careful study of all the relevant factors like land topography, soil, water, crop and agro-climatic conditions, and suitability of drip irrigation system and its components.
- In lighter soils subsurface drip may be unable to wet the soil surface for germination. Requires careful consideration of the installation depth.
- Most drip systems are designed for high efficiency, meaning little or no leaching fraction. Without sufficient leaching, salts applied with the irrigation water may build up in the root

zone, usually at the edge of the wetting pattern. On the other hand, drip irrigation avoids the high capillary potential of traditional surface-applied irrigation, which can draw salt deposits up from deposits below.

- the PVC pipes often suffer from rodent damage, requiring replacement of the entire tube and increasing expenses.
- Drip irrigation systems cannot be used for damage control by night frosts (like in the case of sprinkler irrigation systems).

Uses

Drip irrigation is used in farms, commercial greenhouses, and residential gardeners. Drip irrigation is adopted extensively in areas of acute water scarcity and especially for crops and trees such as coconuts, containerized landscape trees, grapes, bananas, pandey, eggplant, citrus, strawberries, sugarcane, cotton, maize, and potatoes.

Irrigation dripper.

Drip irrigation for garden available in drip kits are increasingly popular for the homeowner and consist of a timer, hose and emitter. Hoses that are 4 mm in diameter are used to irrigate flower pots.

Bioherbicide

A bioherbicide is a biologically based control agent for weeds. Among the three major types of pesticides (agricultural pest-control agents) herbicides are used to control weeds, or undesirable plants. (The other major classes of pesticides are insecticides, which control insects; and fungicides, which control fungi. Normally, any pesticide can only be expected to control, but not eradicate, agricultural pests.)

Almost every agricultural pest has at least one naturally occurring enemy that will reduce its population. Bioherbicides are weed-control agents that use such naturally occurring enemies, rather than depending on man-made chemicals. This can be important because agents of biological control ordinarily have many fewer, and much milder, effects on the environment than do synthetic chemicals. What is more, they tend not to lead to the public health problems that chemicals are associated with. These two advantages of biological control agents, including bioherbicides, do not however mean that they need not also be subjected to careful tests for environmental and public health safety.

Bioherbicides may be compounds derived from microbes such as fungi, bacteria or protozoa; or phytotoxic plant residues, extracts or single compounds derived from other plant species.

A bioherbicide based on a fungus is called a mycoherbicide.

In the industry, bioherbicides and other biopesticides are often referred to as "naturals".

Bush Regeneration

Bush regeneration, a form of natural area restoration, is the term used in Australia for the ecological restoration of remnant vegetation areas, such as through the minimisation of negative disturbances, both exogenous such as exotic weeds and endogenous such as erosion. It may also attempt to recreate conditions of pre-European arrival, for example by simulating endogenous disturbances such as fire. Bush regeneration attempts to protect and enhance the floral biodiversity in an area by providing conditions conducive to the recruitment and survival of native plants. Bushcare's Major Day Out is an Australian national day of community participation in the care of bushland. In 2012 nearly 100 bushcare sites participated in this annual event.

History

Bradley Method

In the early 1960s Joan and Eileen Bradley developed a series of weed control techniques through a process of trial and error. Their work was the beginning of minimal disturbance bush regeneration in New South Wales. The Bradley method urges a naturalistic approach by encouraging the native vegetation to self-reestablish. The Bradleys used their method to successfully clear weeds from a 16 hectares (40 acres) reserve in Ashton Park, part of Sydney Harbour National Park, NSW. The process demonstrated that, following a period of consecutive 'follow up' treatments of diminishing time requirement, subsequent maintenance was needed only once or twice a year, mainly in vulnerable spots such as creek banks, roadsides, and clearings, to be maintained weed-free.

The aim of their work was to clear small niches adjacent to healthy native vegetation such that the each area will regenerate from in-situ soil seed banks or be re-colonised and stabilized by the regeneration of native plants, replacing an area previously occupied by weeds. The Bradley method follows three main principles,

1. secure the best areas first.
2. minimise disturbance to the natural conditions (e.g. minimise soil disturbance and off-target damage).
3. don't overclear, let the regenerative ability of the bush set the pace of clearance (Bradley 1988).

The priority securing of the best quality vegetation aids in preserving areas of top biodiversity which provide regeneration potential to expand these areas and reclaim areas as bushland.

Modern Practice

The adoption of minimal disturbance bush regeneration increased in the decades that followed the work of the Bradleys. Their principles have guided bushcare programs in Australia, although the inclusion of herbicide in modern bush regeneration is a notable deviation from the ideals of the Bradley sisters. In addition, rather than 'minimal disturbance', a more favoured and ecologically sound trend since the 1990s has been towards more 'appropriate disturbance' as many Australian plant communities require some level of perturbation to trigger germination from long-buried seed banks. This has led to a range of additional disturbance-based techniques (such as burns and soil disturbance) being included in the regenerator's 'tool kit' in dry forest and grassland areas. Field experience has found that, even in rainforest areas, a resilience to disturbance is evident, enabling regenerators to clear weed in a fairly extensive manner to trigger rainforest recovery. This is borne out by a thriving rainforest regeneration industry in northern NSW Australia, modelled on the pioneering work of John Stockard at Wingham Brush (Stockard 1991, Stockard 1999). The rule of thumb in all cases is to constrain clearing to that area that matches the project's follow up resources.

The increased awareness and consideration of Australia's biodiversity by citizens has incrementally increased pressure on local governments to adopt conservation programs for remnant vegetation on council land. Most peri urban councils now have some involvement in bush regeneration, either through planning, land management, volunteer support or through employment of bush regeneration practitioners. In NSW the level of coordination of bush regeneration programs through local governments is high, although in some other areas at present a lack of coordination is a serious concern in bush regeneration on public land, with only 40% of councils liaising with other councils. In such areas there may be a need for strategic management at a regional scale through Natural Resource Management Boards or non government organisations such as Trees For Life, which are involved in bushcare programs across wider areas.

Purposes

The aim of bush regeneration, also known as 'natural area restoration', is to restore and maintain ecosystem health by facilitating the natural regeneration of indigenous flora, this is usually achieved by selectively reducing the competitive interaction with invasive species, or mitigation of negative influences such as weeds or erosion.

Invasive plant species are often the greatest threat to remnant vegetation, and therefore bush regeneration is closely associated with weed abatement activities. Weed management as one aim of bush regeneration, is used to increase native plant recruitment. The management of factors such as fire and herbivory can be just as important, depending on the ecosystem being restored. In recent years research and on-ground management has begun to recognize the importance of ecosystem processes rather ecosystem composition and structure and research into other ways of facilitating native plant recruitment is increasing.

Technique

The original Bradley method of bush regeneration focuses on facilitating native plant recruitment from the seedbank, rather than planting seedlings or sowing seeds, as follows:

"Weeding a little at a time from the bush towards the weeds takes the pressure off the natives under favourable conditions. Native seeds and spores are ready in the ground and the natural environment favours plants that have evolved in it. The balance is tipped back towards regeneration. Keep it that way, by always working where the strongest area of bush meets the weakest weeds".

Currently the term 'bush regeneration' includes activities other than weed removal, such as re-planting and introducing species into an area where soil, water, or fire regimes have shifted the type of plant appropriate to the area (e.g. a stormwater drain).

Weed species can be important habitat for native fauna (e.g. blackberry is important habitat for wrens and the southern brown bandicoot) and this should be taken into consideration with bush regeneration, for example by not clearing invasive species until adequate habitat alternatives have been established nearby with native vegetation.

Problems can occur when insufficient follow-up is conducted as the success of bush regeneration is dependent on allowing the native vegetation to regenerate in the area where weeds have been removed.

List of Bushcare Groups

Organisations Offering Community Training in Bush Regeneration

- Trees For Life.
- Campbelltown Council, NSW, Streamcare Group.
- San Diego Chapter, California Native Plant Society.

Reserves where Volunteer Groups Undertake Bush Regeneration

- George Kendall Riverside Park.
- Whites Creek (Annandale).
- Puckeys Estate Reserve.
- Mermaid Pool, Manly Vale Sydney.

Cultivator

F210 Honda tiller.

1949 Farmall C with C-254-A two-row cultivator.

A cultivator pulled by a tractor in Canada in 1943.

A cultivator is any of several types of farm implement used for secondary tillage. One sense of the name refers to frames with the teeth (also called shanks) that pierce the soil as they are dragged through it linearly. Another sense refers to machines that use rotary motion of disks or teeth to accomplish a similar result. The rotary tiller is a principle example.

Cultivators stir and pulverize the soil, either before planting (to aerate the soil and prepare a smooth, loose seedbed) or after the crop has begun growing (to kill weeds—controlled disturbance of the topsoil close to the crop plants kills the surrounding weeds by uprooting them, burying their leaves to disrupt their photosynthesis, or a combination of both). Unlike a harrow, which disturbs the entire surface of the soil, cultivators are designed to disturb the soil in careful patterns, sparing the crop plants but disrupting the weeds.

Cultivators of the toothed type are often similar in form to chisel plows, but their goals are different. Cultivator teeth work near the surface, usually for weed control, whereas chisel plow shanks work deep beneath the surface, breaking up hardpan. Consequently, cultivating also takes much less power per shank than does chisel plowing.

Small toothed cultivators pushed or pulled by a single person are used as garden tools for small-scale gardening, such as for the household's own use or for small market gardens. Similarly sized rotary tillers combine the functions of harrow and cultivator into one multipurpose machine.

Cultivators are usually either self-propelled or drawn as an attachment behind either a two-wheel tractor or four-wheel tractor. For two-wheel tractors they are usually rigidly fixed and powered via couplings to the tractors' transmission. For four-wheel tractors they are usually attached by means of a three-point hitch and driven by a power take-off (PTO). Drawbar hookup is also still commonly used worldwide. Draft-animal power is sometimes still used today, being somewhat common in developing nations although rare in more industrialized economies.

History

The basic idea of soil scratching for weed control is ancient and was done with hoes or mattocks for millennia before cultivators were developed. Cultivators were originally drawn by draft animals (such as horses, mules, or oxen) or were pushed or drawn by people. In modern commercial agriculture, the amount of cultivating done for weed control has been greatly reduced via use of herbicides instead. However, herbicides are not always desirable—for example, in organic farming.

The powered rotary hoe was invented by Arthur Clifford Howard who, in 1912, began experimenting with rotary tillage on his father's farm at Gilgandra, New South Wales, Australia. Initially using his father's steam tractor engine as a power source, he found that ground could be mechanically tilled without soil-packing occurring, as was the case with normal ploughing. His earliest designs threw the tilled soil sideways, until he improved his invention by designing an L-shaped blade mounted on widely spaced flanges fixed to a small-diameter rotor. With fellow apprentice Everard McCleary, he established a company to make his machine, but plans were interrupted by World War I. In 1919 Howard returned to Australia and resumed his design work, patenting a design with 5 rotary hoe cultivator blades and an internal combustion engine in 1920.

In March 1922, Howard formed the company Austral Auto Cultivators Pty Ltd, which later became known as Howard Auto Cultivators. It was based in Northmead, a suburb of Sydney, from 1927.

Meanwhile, in North America during the 1910s, tractors were evolving away from traction engine-sized monsters toward smaller, lighter, more affordable machines. The Fordson tractor especially had made tractors affordable and practical for small and medium family farms for the first time in history. Cultivating was somewhat of an afterthought in the Fordson's design, which reflected the fact that even just bringing practical motorized tractive power alone to this market segment was in itself a milestone. This left an opportunity for others to pursue better motorized cultivating. Between 1915 and 1920, various inventors and farm implement companies experimented with a class of machines referred to as *motor cultivators*, which were simply modified horse-drawn shank-type cultivators with motors added for self-propulsion. This class of machines found limited market success. But by 1921 International Harvester had combined motorized cultivating with the other tasks of tractors (tractive power and belt work) to create the Farmall, the general-purpose tractor tailored to cultivating that basically invented the category of row-crop tractors.

In Australia, by the 1930s, Howard was finding it increasingly difficult to meet a growing worldwide demand for exports of his machines. He travelled to the United Kingdom, founding the company Rotary Hoes Ltd in East Horndon, Essex, in July 1938. Branches of this new company subsequently opened in the United States of America, South Africa, Germany, France, Italy, Spain, Brazil, Malaysia, Australia and New Zealand. It later became the holding company for Howard Rotavator Co. Ltd. The Howard Group of companies was acquired by the Danish Thrige Agro Group in 1985, and in December 2000 the Howard Group became a member of Kongskilde Industries of Soroe, Denmark.

When herbicidal weed control was first widely commercialized in the 1950s and 1960s, it played into that era's optimistic worldview in which sciences such as chemistry would usher in a new age of modernity that would leave old-fashioned practices (such as weed control via cultivators) in the dustbin of history. Thus herbicidal weed control was adopted very widely, and in some cases too heavily and hastily. In subsequent decades, people overcame this initial imbalance and came to realize that herbicidal weed control has limitations and externalities, and it must be managed intelligently. It is still widely used, and probably will continue to be indispensable to affordable food production worldwide for the foreseeable future; but its wise management includes seeking alternate methods, such as the traditional standby of mechanical cultivation, where practical.

Industrial Use

To the extent that cultivating is done commercially today (such as in truck farming), it is usually powered by tractors, especially row-crop tractors. Industrial cultivators can vary greatly in size and shape, from 10 feet (3 m) to 80 feet (24 m) wide. Many are equipped with hydraulic wings that fold up to make road travel easier and safer. Different types are used for preparation of fields before planting, and for the control of weeds between row crops. The cultivator may be an implement trailed after the tractor via a drawbar; mounted on the three-point hitch; or mounted on a frame beneath the tractor. Active cultivator implements are driven by a power take-off shaft. While most cultivator are considered a secondary tillage implement, active cultivators are commonly used for primary tillage in lighter soils instead of plowing. The largest versions available are about 6 m (20 ft) wide, and require a tractor with an excess of 150 horsepower (110 kW) (PTO) to drive them.

Field cultivators are used to complete tillage operations in many types of arable crop fields. The main function of the field cultivator is to prepare a proper seedbed for the crop to be planted into, to bury crop residue in the soil (helping to warm the soil before planting), to control weeds, and to mix and incorporate the soil to ensure the growing crop has enough water and nutrients to grow well during the growing season. The implement has many shanks mounted on the underside of a metal frame, and small narrow rods at the rear of the machine that smooth out the soil surface for easier travel later when planting. In most field cultivators, one-to-many hydraulic cylinders raise and lower the implement and control its depth.

Row Crop Cultivators

The main function of the row crop cultivator is weed control between the rows of an established crop. Row crop cultivators are usually raised and lowered by a three-point hitch and the depth is controlled by gauge wheels.

Home made sweep. Notice the inner and outer "sweep" blades.

Sometimes referred to as *sweep cultivators*, these commonly have two center blades that cut weeds from the roots near the base of the crop and turn over soil, while two rear sweeps further outward than the center blades deal with the center of the row, and can be anywhere from 1 to 36 rows wide.

Garden Cultivators

Small tilling equipment, used in small gardens such as household gardens and small commercial gardens, can provide both primary and secondary tillage. For example, a rotary tiller does both the "plowing" and the "harrowing", preparing a smooth, loose seedbed. It does not provide the row-wise weed control that cultivator teeth would. For that task, there are single-person-pushable toothed cultivators.

Variants and Trademarks

A Japanese two-wheel tractor.

Rotary tillers are a type of cultivators. Rotary tillers are popular with home gardeners who want large vegetable gardens. The garden may be tilled a few times before planting each crop. Rotary tillers may be rented from tool rental centers for single-use applications, such as when planting grass.

A small rotary hoe for domestic gardens was known by the trademark Rototiller and another, made by the Howard Group, who produced a range of rotary tillers, was known as the Rotavator.

Rototiller:

The small rototiller is typically propelled forward via a (1–5 horsepower or 0.8–3.5 kilowatts) petrol engine rotating the tines, and do not have powered wheels, though they may have small transport/level control wheel(s). To keep the machine from moving forward too fast, an adjustable tine is usually fixed just behind the blades so that through friction with deeper un-tilled soil, it acts as a brake, slowing the machine and allowing it to pulverize the soils. The slower a rototiller moves forward, the more soil tilth can be obtained. The operator can control the amount of friction/braking action by raising and lowering the handlebars of the tiller. Rototillers do not have a reverse as such backwards movement towards the operator could cause serious injury. While operating, the rototiller can be pulled backwards to go over areas that were not pulverized enough, but care must be taken to ensure that the operator does not stumble and pull the rototiller on top of himself. Rototilling is much faster than manual tilling, but notoriously difficult to handle and exhausting work, especially in the heavier and higher horsepower models. If the rototiller's blades catch on unseen subsurface objects, such as tree roots and buried garbage, it can cause the rototiller to abruptly and violently move in any direction.

Rotavator:

Unlike the Rototiller, the self-propelled Howard Rotavator is equipped with a gearbox and driven forward, or held back, by its wheels. The gearbox enables the forward speed to be adjusted while the rotational speed of the tines remains constant which enables the operator to easily regulate the extent to which soil is engaged. For a two-wheel tractor rotavator this greatly reduces the workload of the operator as compared to a rototiller. These rotavators are generally more heavy duty, come in higher power (4–18 horsepower or 3–13 kilowatts) with either petrol or diesel engines and can cover much more area per hour. The trademarked word "Rotavator" is one of the longest single-word palindromes in the English language.

Mini tiller:

Mini tillers are a new type of small agricultural tillers or cultivators used by farmers or homeowners. These are also known as power tillers or garden tillers. Compact, powerful and, most importantly, inexpensive, these agricultural rotary tillers are providing alternatives to four-wheel tractors and in the small farmers' fields in developing countries are more economical than four-wheel tractors.

Two-wheel tractor:

The higher power "riding" rotavators cross out of the home garden category into farming category, especially in Asia, Africa and South America, capable of preparing 1 hectare of land in 8–10 hours. These are also known as *power tillers* or *walking tractors*. Years ago they were considered only useful for rice growing areas, where they were fitted with steel cage-wheels for traction, but now the same are being used in both wetland and dryland farming all over the world. They have multiple functions with related tools for dryland or paddys, pumping, transportation, threshing, ditching, spraying pesticide. They can be used on hills, mountains, in greenhouses and orchards. Diesel designs are more popular in developing countries than gasoline.

Mulch

A mulch is a layer of material applied to the surface of an area of soil. Its purpose is any or all of the following:

- To conserve moisture.
- To improve the fertility and health of the soil.
- To reduce weed growth.
- To enhance the visual appeal of the area.

A mulch is usually but not exclusively organic in nature. It may be permanent (e.g. plastic sheeting) or temporary (e.g. bark chips). It may be applied to bare soil, or around existing plants. Mulches of manure or compost will be incorporated naturally into the soil by the activity of worms and other organisms. The process is used both in commercial crop production and in gardening, and when applied correctly can dramatically improve soil productivity.

Uses

Many materials are used as mulches, which are used to retain soil moisture, regulate soil temperature, suppress weed growth, and for aesthetics. They are applied to the soil surface, around trees, paths, flower beds, to prevent soil erosion on slopes, and in production areas for flower and vegetable crops. Mulch layers are normally two inches or more deep when applied.

They are applied at various times of the year depending on the purpose. Towards the beginning of the growing season mulches serve initially to warm the soil by helping it retain heat which is lost during the night. This allows early seeding and transplanting of certain crops, and encourages faster growth. As the season progresses, mulch stabilizes the soil temperature and moisture, and prevents the growing of weeds from seeds. In temperate climates, the effect of mulch is dependent upon the time of year they are applied and when applied in fall and winter, are used to delay the growth of perennial plants in the spring or prevent growth in winter during warm spells, which limits freeze thaw damage.

The effect of mulch upon soil moisture content is complex. Mulch forms a layer between the soil and the atmosphere which prevents sunlight from reaching the soil surface, thus reducing evaporation. However, mulch can also prevent water from reaching the soil by absorbing or blocking water from light rains.

In order to maximise the benefits of mulch, while minimizing its negative influences, it is often applied in late spring/early summer when soil temperatures have risen sufficiently, but soil moisture content is still relatively high. However, permanent mulch is also widely used and valued for its simplicity, as popularized by author Ruth Stout, who said, “My way is simply to keep a thick mulch of any vegetable matter that rots on both sides of my vegetable and flower garden all year long. As it decays and enriches the soils, I add more.”

Plastic mulch used in large-scale commercial production is laid down with a tractor-drawn or standalone layer of plastic mulch. This is usually part of a sophisticated mechanical process, where raised beds are formed, plastic is rolled out on top, and seedlings are transplanted through it. Drip irrigation is often required, with drip tape laid under the plastic, as plastic mulch is impermeable to water.

Materials

Rubber mulch nuggets in a playground. The white fibers are nylon cords, which are present in the tires from which the mulch is made.

Shredded wood used as mulch. This type of mulch is often dyed to improve its appearance in the landscape.

Spring daffodils push through shredded wood mulch.

Materials used as mulches vary and depend on a number of factors. Use takes into consideration availability, cost, appearance, the effect it has on the soil—including chemical reactions and pH, durability, combustibility, rate of decomposition, how clean it is—some can contain weed seeds or plant pathogens.

A variety of materials are used as mulch:

- Organic residues: grass clippings, leaves, hay, straw, kitchen scraps comfrey, shredded bark, whole bark nuggets, sawdust, shells, woodchips, shredded newspaper, cardboard, wool, animal manure, etc. Many of these materials also act as a direct composting system, such as the mulched clippings of a mulching lawn mower, or other organics applied as sheet composting.
- Compost: fully composted materials are used to avoid possible phytotoxicity problems. Materials that are free of seeds are ideally used, to prevent weeds being introduced by the mulch.
- Old carpet (synthetic or natural): makes a free, readily available mulch.
- Rubber mulch: made from recycled tire rubber.
- Plastic mulch: crops grow through slits or holes in thin plastic sheeting. This method is predominant in large-scale vegetable growing, with millions of acres cultivated under plastic mulch worldwide each year (disposal of plastic mulch is cited as an environmental problem).
- Rock and gravel can also be used as a mulch. In cooler climates the heat retained by rocks may extend the growing season.

In some areas of the United States, such as central Pennsylvania and northern California, mulch is often referred to as "tanbark", even by manufacturers and distributors. In these areas, the word "mulch" is used specifically to refer to very fine tanbark or peat moss.

Organic Mulches

Organic mulches decay over time and are temporary. The way a particular organic mulch decomposes and reacts to wetting by rain and dew affects its usefulness.

Some mulches such as straw, peat, sawdust and other wood products may for a while negatively affect plant growth because of their wide carbon to nitrogen ratio, because bacteria and fungi that decompose the materials remove nitrogen from the surrounding soil for growth. However, whether this effect has any practical impact on gardens is disputed by researchers and the experience of gardeners. Organic mulches can mat down, forming a barrier that blocks water and air flow between the soil and the atmosphere. Vertically applied organic mulches can wick water from the soil to the surface, which can dry out the soil. Mulch made with wood can contain or feed termites, so care must be taken about not placing mulch too close to houses or building that can be damaged by those insects. Some mulch manufacturers recommend putting mulch several inches away from buildings.

Commonly available organic mulches include:

Leaves

- *Leaves* from deciduous trees, which drop their foliage in the autumn/fall. They tend to be dry and blow around in the wind, so are often chopped or shredded before application. As they decompose they adhere to each other but also allow water and moisture to seep down to the soil surface. Thick layers of entire leaves, especially of maples and oaks, can form a soggy mat in winter and spring which can impede the new growth lawn grass and other plants. Dry leaves are used as winter mulches to protect plants from freezing and thawing in areas with cold winters, they are normally removed during spring.

Grass Clippings

- *Grass clippings*, from mowed lawns are sometimes collected and used elsewhere as mulch. Grass clippings are dense and tend to mat down, so are mixed with tree leaves or rough compost to provide aeration and to facilitate their decomposition without smelly putrefaction. Rotting fresh grass clippings can damage plants; their rotting often produces a damaging buildup of trapped heat. Grass clippings are often dried thoroughly before application, which mediates against rapid decomposition and excessive heat generation. Fresh green grass clippings are relatively high in nitrate content, and when used as a mulch, much of the nitrate is returned to the soil, but the routine removal of grass clippings from the lawn results in nitrogen deficiency for the lawn.

Peat Moss

- *Peat moss*, or sphagnum peat, is long lasting and packaged, making it convenient and popular as a mulch. When wetted and dried, it can form a dense crust that does not allow water to soak in. When dry it can also burn, producing a smoldering fire. It is sometimes mixed with pine needles to produce a mulch that is friable. It can also lower the pH of the soil surface, making it useful as a mulch under acid loving plants.

However peat bogs are a valuable wildlife habitat, and peat is also one of the largest stores of carbon (in Britain, out of a total estimated 9952 million tonnes of carbon in British vegetation and soils, 6948 million tonnes carbon are estimated to be in Scottish, mostly peatland, soils), so gardeners who wish to protect the environment will choose more sustainable alternatives.

Wood Chips

- *Wood chips* are a byproduct of the pruning of trees by arborists, utilities and parks; they are used to dispose of bulky waste. Tree branches and large stems are rather coarse after chipping and tend to be used as a mulch at least three inches thick. The chips are used to conserve soil moisture, moderate soil temperature and suppress weed growth. The decay of freshly produced chips from recently living woody plants, consumes nitrate; this is often off set with a light application of a high-nitrate fertilizer. Wood chips are most often used under trees and shrubs. When used around soft stemmed plants, an unmulched zone is left around the plant stems to prevent stem rot or other possible diseases. They are often used to mulch trails, because they are readily produced with little additional cost outside of the normal disposal cost of tree maintenance. Wood chips come in various colors.

Woodchip Mulch

- *Woodchip mulch* is a byproduct of reprocessing used (untreated) timber (usually packaging pallets), to dispose of wood waste by creating woodchip mulch. The chips are used to conserve soil moisture, moderate soil temperature and suppress weed growth. Woodchip mulch is often used under trees, shrubs or large planting areas and can last much longer than arborist mulch. In addition, many consider woodchip mulch to be visually appealing, as it comes in various colors. Woodchips can also be reprocessed into playground woodchip to be used as an impact-attenuating playground surfacing.

Bark Chips

Bark chips.

- *Bark chips* of various grades are produced from the outer corky bark layer of timber trees. Sizes vary from thin shredded strands to large coarse blocks. The finer types are very attractive but have a large exposed surface area that leads to quicker decay. Layers two or three inches deep are usually used, bark is relativity inert and its decay does not demand soil nitrates. Bark chips are also available in various colors.

Straw Mulch / Field Hay / Salt Hay

Permaculture garden with a fruit tree, herbs, flowers and vegetables mulched with hay.

- *Straw mulch* or *field hay* or *salt hay* are lightweight and normally sold in compressed bales. They have an unkempt look and are used in vegetable gardens and as a winter covering. They are biodegradable and neutral in pH. They have good moisture retention and weed controlling properties but also are more likely to be contaminated with weed seeds. Salt hay is less likely to have weed seeds than field hay. Straw mulch is also available in various colors.

Cardboard / Newspaper

- *Cardboard* or *newspaper* can be used as mulches. These are best used as a base layer upon which a heavier mulch such as compost is placed to prevent the lighter cardboard/newspaper layer from blowing away. By incorporating a layer of cardboard/newspaper into a mulch, the quantity of heavier mulch can be reduced, whilst improving the weed suppressant and moisture retaining properties of the mulch. However, additional labour is expended when planting through a mulch containing a cardboard/newspaper layer, as holes must be cut for each plant. Sowing seed through mulches containing a cardboard/newspaper layer is impractical. Application of newspaper mulch in windy weather can be facilitated by briefly pre-soaking the newspaper in water to increase its weight.

Carpet

- Synthetic carpet that is composed of artificial fibers may be removed after planting to prevent fibers taking a long time to decompose, whereas carpet made from natural fibers may be kept in place, blocking competition from weeds. Rain is absorbed by carpet and then slowly released into the soil, reducing watering needs.

Colored Mulch

Some organic mulches are colored red, brown, black, and other colors. Isopropanolamine, specifically 1-Amino-2-propanol or DOW™ monoisopropanolamine, may be used as a pigment dispersant and color fastener in these mulches. Types of mulch which can be dyed include: wood chips, bark chips (barkdust) and pine straw. Colored mulch is made by dyeing the mulch in a water-based solution of colorant and chemical binder. When colored mulch first entered the market, most formulas were suspected to contain toxic, heavy metals and other contaminates. Today, "current investigations indicate that mulch colorants pose no threat to people, pets or the environment. The dyes currently used by the mulch and soil industry are similar to those used in the cosmetic and other manufacturing industries (i.e., iron oxide)," as stated by the Mulch and Soil Council. Colored mulch can be applied anywhere non-colored mulch is used (such as large bedded areas or around plants) and features many of the same gardening benefits as traditional mulch, such as improving soil productivity and retaining moisture. As mulch decomposes, just as with non-colored mulch, more mulch may need to be added to continue providing benefits to the soil and plants. However, if mulch is faded, spraying dye to previously spread mulch in order to restore color is an option.

Anaerobic (Sour) Mulch

Mulch normally smells like freshly cut wood, but sometimes develops a toxicity that causes it to smell like vinegar, ammonia, sulfur or silage. This happens when material with ample nitrogen content is not rotated often enough and it forms pockets of increased decomposition. When this occurs, the process may become anaerobic and produce these phytotoxic materials in small quantities. Once exposed to the air, the process quickly reverts to an aerobic process, but these toxic materials may be present for a period of time. If the mulch is placed around plants before the toxicity has had a chance to dissipate, then the plants could very likely be damaged or killed depending on their hardiness. Plants that are predominantly low to the ground or freshly planted are the most susceptible, and the phytotoxicity may prevent germination of some seeds.

If sour mulch is applied and there is plant kill, the best thing to do is to water the mulch heavily. Water dissipates the chemicals faster and refreshes the plants. Removing the offending mulch may have little effect, because by the time plant kill is noticed, most of the toxicity is already dissipated. While testing after plant kill will not likely turn up anything, a simple pH check may reveal high acidity, in the range of 3.8 to 5.6 instead of the normal range of 6.0 to 7.2. Finally, placing a bit of the offending mulch around another plant to check for plant kill will verify if the toxicity has departed. If the new plant is also killed, then sour mulch is probably not the problem.

Groundcovers (Living Mulches)

Groundcovers are plants which grow close to the ground, under the main crop, to slow the development of weeds and provide other benefits of mulch. They are usually fast-growing plants that continue growing with the main crops. By contrast, cover crops are incorporated into the soil or killed with herbicides. However, live mulches also may need to be mechanically or chemically killed eventually to prevent competition with the main crop.

Some groundcovers can perform additional roles in the garden such as nitrogen fixation in the case of clovers, dynamic accumulation of nutrients from the subsoil in the case of creeping comfrey (*Symphytum ibericum*), and even food production in the case of *Rubus tricolor*.

On-site Mulch Production

Owing to the great bulk of mulch which is often required on a site, it is often impractical and expensive to source and import sufficient mulch materials. An alternative to importing mulch materials is to grow them on site in a "mulch garden" - an area of the site dedicated entirely to the production of mulch which is then transferred to the growing area. Mulch gardens should be sited as close as possible to the growing area so as to facilitate transfer of mulch materials.

Mulching (Composting) Over Unwanted Plants

Sufficient mulch over plants will destroy them, and may be more advantageous than using herbicide, cutting, mowing, pulling, raking, or tilling. The higher the temperature that this "mulch" is composted, the quicker the reduction of undesirable materials. "Undesirable materials" may include living seed, plant "trash", as well as pathogens such as from animal feces, urine (e.g. hantavirus), fleas, lice, ticks, etc.

In some ways this improves the soil by attracting and feeding earthworms, and adding humus. Earthworms "till" the soil, and their feces are among the best fertilizers and soil conditioners.

Urine may be toxic to plants if applied to growing areas undiluted.

Polypropylene and Polyethylene Mulch

Polypropylene mulch is made up of polypropylene polymers where polyethylene mulch is made up of polyethylene polymers. These mulches are commonly used in many plastics. Polyethylene is used mainly for weed reduction, where polypropylene is used mainly on perennials. This mulch is placed on top of the soil and can be done by machine or hand with pegs to keep the mulch tight against the soil. This mulch can prevent soil erosion, reduce weeding, conserve soil moisture, and increase temperature of the soil. Ultimately this can reduce the amount of work a farmer may have to do, and the amount of herbicides applied during the growing period. The black and clear mulches capture sunlight and warm the soil increasing the growth rate. White and other reflective colours will also warm the soil, but they do not suppress weeds as well. This mulch may require other sources of obtaining water such as drip irrigation since it can reduce the amount of water that reaches the soil. This mulch needs to be manually removed at the end of the season since when it starts to break down it breaks down into smaller pieces. If the mulch is not removed before it starts to break down eventually it will break down into ketones and aldehydes polluting the soil. This mulch is technically biodegradable but does not break down into the same materials the more natural biodegradable mulch does.

Biodegradable Mulch

Quality biodegradable mulches are made out of plant starches and sugars or polyester fibres. These starches can come from plants such as wheat and corn. These mulch films may be a bit more permeable allowing more water into the soil. This mulch can prevent soil erosion, reduce weeding, conserve soil moisture, and increase temperature of the soil. Ultimately this can reduce the amount of herbicides used and manual labour farmers may have to do throughout the growing season. At the end of the season these mulches will start to break down from heat. Microorganisms in the soil break down the mulch into two components, water and CO_2, leaving no toxic residues behind. This source of mulch is even less manual labour since it does not need to be removed at the end of the season and can actually be tilled into the soil. With this mulch its important to take into consideration that its mulch is more delicate then other kinds. It should be placed on a day which is not too hot and with less tension then other synthetic mulches. These also can be placed by machine or hand but its ideal to have a more starchy mulch that will allow it to stick to the soil better.

Hoe Tool

A farmer using a hoe to keep weeds down in a vegetable garden.

A hoe is an ancient and versatile agricultural hand tool used to shape the soil, control weeds, clear soil, and harvest root crops. Shaping the soil can be piling soil around the base of plants (hilling), creating narrow furrows (drills) and shallow trenches for planting seeds and bulbs. Weed control with a hoe can be by agitating the surface of the soil or by cutting foliage from the roots, and clearing soil of old roots and crop residues. Hoes for digging and moving soil are used harvesting root crops such as potatoes.

Types

There are many types of hoes of quite different appearances and purposes. Some can perform multiple functions. Others are intended for a specific use.

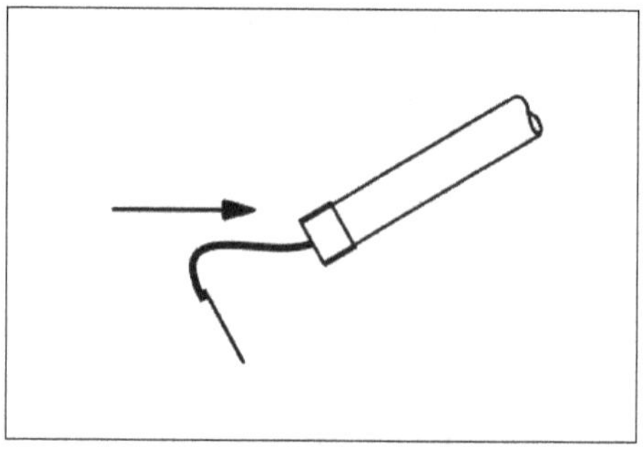

Cultivating tool pull or draw hoe.

There are two main classes of agricultural hoe: draw hoes for shaping, and scuffle hoes for surface weed control.

A draw hoe has the blade set at approximately a right angle to the handle. The user chops into the ground and then pulls (draws) the blade towards them. Altering the angle of the handle can cause the hoe to dig deeper or more shallowly as the hoe is pulled. A draw hoe can easily be used to cultivate soil to a depth of several inches. A typical design of draw hoe, the "eye hoe", has a ring in the head through which the handle is fitted. This design has been used since Roman times.

A scuffle hoe is used to scrape the surface of the soil, and to loosen the top inch or so, and to cut the roots, remove, and disrupt the growth of weeds efficiently. These are mainly of two different designs: the Dutch hoe and the hoop hoe.

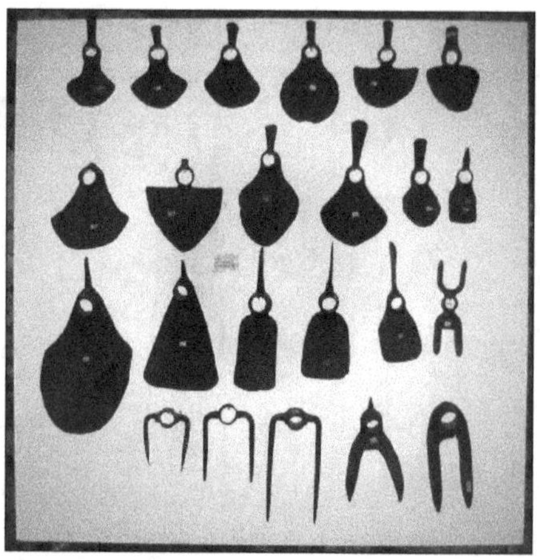

Eye hoe heads, some with sow-tooth (German: Sauzahn), *Centro Etnográfico de Soutelo de Montes*, Pontevedra, Spain.

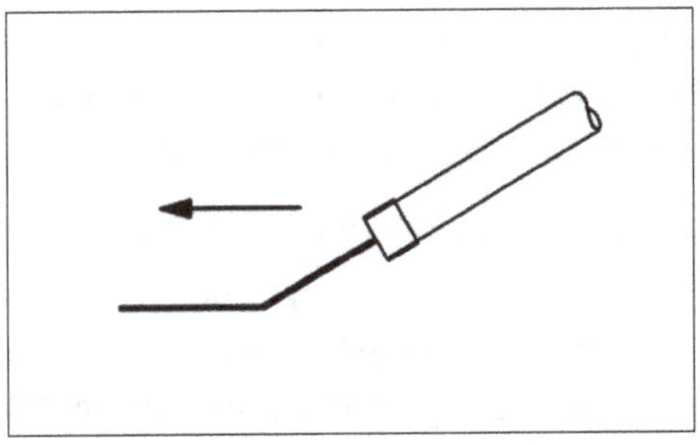

Cultivating tool push or thrust hoe.

The term 'hand hoe' most commonly refers to any type of light-weight, short-handled hoe, although it may be used simply to contrast hand-held tools against animal or machine pulled tools.

Draw Hoes

- The typical farming and gardening hoe with a heavy, broad blade and a straight edge is known as the grub hoe, grab hoe, pattern hoe, Italian hoe, azada or dago hoe ("dago" is an ethnic slur referring to Italians, Spaniards, or Portuguese).
- Ridging hoes also known as warren or drill hoes are triangular (point-down) or heart-shaped draw hoes particularly for creating narrow furrows (drills) and shallow trenches for planting seeds and bulbs.

Hoedad (tree-planting tool) Kaibab National Forest, Arizona, USA.

- Hoedads (also, "hoedags") are hoe-like tools used for planting trees. According to Hartzell (1987, p. 29), "The hoedag [was] originally called skindvic hoe... Hans Rasmussen, legendary contractor and timber farm owner, is credited with having invented the curved, convex, round-nosed hoedag blade which is widely used today" (emphasis added).
- The mortar hoe, a tool specific to hand mixing mortar and concrete, has the appearance of a typical square-bladed draw hoe with the addition of large holes in the blade.

Scuffle Hoes

- The Dutch hoe is a design that is pushed or pulled through the soil to cut weeds just under the surface. A Dutch hoe has a blade "sharp on every side so as to cut either forwards and backwards". The blade must be set in a plane slightly upwardly inclined to the dual axis of the rod used as a handle stick. The user uses the handle to push the blade forward, forcing it below the surface of the ground and maintaining it at a shallow depth in the surface layer of soil by altering the angle of the handle whilst pushing. A push hoe can easily cultivate and remove weeds etc. from the surface layer of the soil.

- The hoop hoe (also known as action, oscillating, hula, stirrup, pendulum weeder, or swivel hoes) have a double-edge blade that bends around to form a rectangle attached to the handle. Weeds are cut just below the soil surface as the blade is pushed & pulled through the area. The back and forth motion is highly effective with cutting weeds in loose or breakable soil. Widths of the blade typically range between three and seven inches. Its tool-head is a loop of flat, sharpened strap metal. It is not as efficient as a draw hoe for moving soil.
- The collinear hoe has a narrow, razor-sharp blade which is used to slice weeds by skimming it just under the surface of the soil with a sweeping motion; it is unsuitable for tasks like soil moving and chopping. It was designed by Eliot Coleman in the late 1980s.
- The Swoe hoe is a modern one-sided cutting hoe - a variant of the Dutch hoe.

Other Hoes

Hoes resembling neither draw nor scuffle hoes include:

- Wheel hoes are, as the name suggests, a hoe or pair of hoes attached to one or more wheels. The hoes are frequently interchangeable with other tools.
- Horse hoes, resembling small ploughs, were a favourite implement of agricultural pioneer Jethro Tull, claiming in his book "Horse Hoeing Husbandry" that "the horse-hoe will, in wide intervals, give wheat throughout all the stages of its life, as much nourishment as the discreet hoer pleases". The modern view is that, rather than nutrients being released, the crop simply benefits from the removal of competing plants. The introduction of the horse hoe, together with the better-known seed drill, brought about the great increase farming productivity seen during the British Agricultural Revolution.

- Fork hoes, (also known as prong hoes, tined hoes or bent forks) are hoes that have two or more tines at right angles to the shaft. Their use is typically to loosen the soil, prior to planting or sowing.

- Clam hoes, made for clam digging.
- Adze hoes, with the basic hoe shape but heavier and stronger and with traditional uses in trail making.
- Pacul or cangkul (hoes similar to adze hoe from Malaysia and Indonesia).
- Gang hoes for powered use (in use at least from 1887 to 1964).

History

Hoes are an ancient technology, predating the plough and perhaps preceded only by the digging stick. In Sumerian mythology, the invention of the hoe was credited to Enlil, the chief of the council of gods. The hand-plough (*mr*) was depicted in predynastic Egyptian art, and hoes are also mentioned in ancient documents like the Code of Hammurabi (ca. 18th century BC) and the Book of Isaiah (c. 8th century BC).

The human damage caused by long-term use of short-handled hoes, which required the user to bend over from the waist to reach the ground, and caused permanent, crippling lower back pain to farm workers, resulted, after struggle led by César Chávez with political help from Governor Jerry Brown in the California Supreme Court declaring the short-handled hoe to be an unsafe hand tool that was banned under California law in 1975.

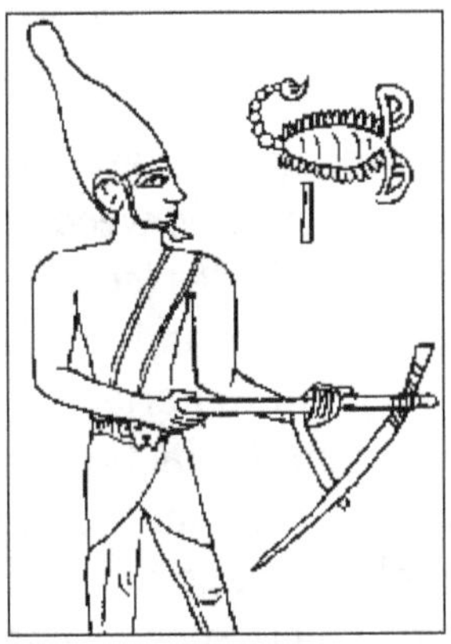

'Mr' hand-plough, Protodynastic Period of Egypt (from the Scorpion Macehead).

- Ancient Greek iron hoe (Kerameikos Archaeological Museum).

A 2000-year-old iron Roman hoe blade.

A Dutch hoe or push hoe; usually attached to a long hilt and handle.

Archaeological Use

Over the past fifteen or twenty years, hoes have become increasingly popular tools for professional archaeologists. While not as accurate as the traditional trowel, the hoe is an ideal tool for cleaning relatively large open areas of archaeological interest. It is faster to use than a trowel, and produces a much cleaner surface than an excavator bucket or shovel-scrape, and consequently on many open-area excavations the once-common line of kneeling archaeologists trowelling backwards has been replaced with a line of stooping archaeologists with hoes.

Herbicide

Herbicide(s), also commonly known as weedkillers, are chemical substances used to control unwanted plants. Selective herbicides control specific weed species, while leaving the desired crop relatively unharmed, while non-selective herbicides (sometimes called "total weedkillers" in commercial products) can be used to clear waste ground, industrial and construction sites, railways and railway embankments as they kill all plant material with which they come into contact. Apart from selective/non-selective, other important distinctions include *persistence* (also known as *residual action*: how long the product stays in place and remains active), *means of uptake* (whether it is absorbed by above-ground foliage only, through the roots, or by other means), and *mechanism of action* (how it works). Historically, products such as common salt and other metal salts were used as herbicides, however these have gradually fallen out of favor and in some countries a number of these are banned due to their persistence in soil, and toxicity and groundwater contamination concerns. Herbicides have also been used in warfare and conflict.

Weeds controlled with herbicide.

Modern herbicides are often synthetic mimics of natural plant hormones which interfere with growth of the target plants. The term organic herbicide has come to mean herbicides intended for organic farming; these are often less efficient and more costly than synthetic herbicides and are based on natural materials. Some plants also produce their own natural herbicides, such as the genus *Juglans* (walnuts), or the tree of heaven; such action of natural herbicides, and other related chemical interactions, is called allelopathy. Due to herbicide resistance - a major concern in agriculture - a number of products also combine herbicides with different means of action.

In the US in 2007, about 83% of all herbicide usage, determined by weight applied, was in agriculture. In 2007, world pesticide expenditures totaled about $39.4 billion; herbicides were about 40% of those sales and constituted the biggest portion, followed by insecticides, fungicides, and other types. Smaller quantities are used in forestry, pasture systems, and management of areas set aside as wildlife habitat.

History

Prior to the widespread use of chemical herbicides, cultural controls, such as altering soil pH, salinity, or fertility levels, were used to control weeds. Mechanical control (including tillage) was also (and still is) used to control weeds.

First Herbicides

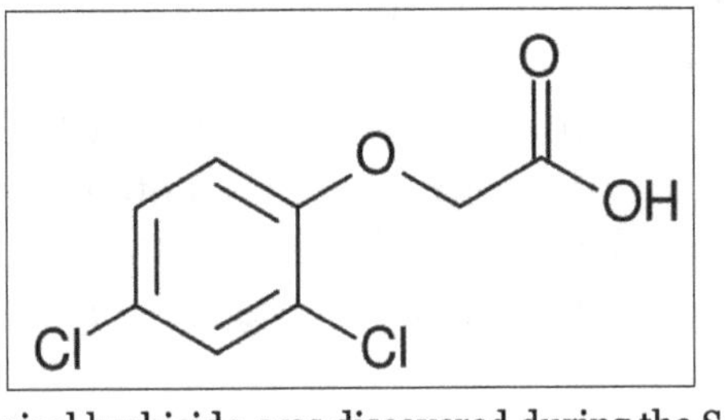

2,4-D, the first chemical herbicide, was discovered during the Second World War.

Although research into chemical herbicides began in the early 20th century, the first major breakthrough was the result of research conducted in both the UK and the US during the Second World War into the potential use of agents as biological weapons. The first modern herbicide, 2,4-D, was first discovered and synthesized by W. G. Templeman at Imperial Chemical Industries. In 1940, he showed that "Growth substances applied appropriately would kill certain broad-leaved weeds in cereals without harming the crops." By 1941, his team succeeded in synthesizing the chemical. In the same year, Pokorny in the US achieved this as well.

Independently, a team under Juda Hirsch Quastel, working at the Rothamsted Experimental Station made the same discovery. Quastel was tasked by the Agricultural Research Council (ARC) to discover methods for improving crop yield. By analyzing soil as a dynamic system, rather than an inert substance, he was able to apply techniques such as perfusion. Quastel was able to quantify the influence of various plant hormones, inhibitors and other chemicals on the activity of microorganisms in the soil and assess their direct impact on plant growth. While the full work of the unit remained secret, certain discoveries were developed for commercial use after the war, including the 2,4-D compound.

When it was commercially released in 1946, it triggered a worldwide revolution in agricultural output and became the first successful selective herbicide. It allowed for greatly enhanced weed control in wheat, maize (corn), rice, and similar cereal grass crops, because it kills dicots (broadleaf plants), but not most monocots (grasses). The low cost of 2,4-D has led to continued usage today, and it remains one of the most commonly used herbicides in the world. Like other acid herbicides, current formulations use either an amine salt (often trimethylamine) or one of many esters of the parent compound. These are easier to handle than the acid.

Further Discoveries

The triazine family of herbicides, which includes atrazine, were introduced in the 1950s; they have the current distinction of being the herbicide family of greatest concern regarding groundwater contamination. Atrazine does not break down readily (within a few weeks) after being applied to soils of above neutral pH. Under alkaline soil conditions, atrazine may be carried into the soil profile as far as the water table by soil water following rainfall causing the aforementioned contamination. Atrazine is thus said to have "carryover", a generally undesirable property for herbicides.

Glyphosate (Roundup) was introduced in 1974 for nonselective weed control. Following the development of glyphosate-resistant crop plants, it is now used very extensively for selective weed control in growing crops. The pairing of the herbicide with the resistant seed contributed to the consolidation of the seed and chemistry industry in the late 1990s.

Many modern chemical herbicides used in agriculture and gardening are specifically formulated to decompose within a short period after application. This is desirable, as it allows crops and plants to be planted afterwards, which could otherwise be affected by the herbicide. However, herbicides with low residual activity (i.e., that decompose quickly) often do not provide season-long weed control and do not ensure that weed roots are killed beneath construction and paving (and cannot emerge destructively in years to come), therefore there remains a role for weedkiller with high levels of persistence in the soil.

Terminology

Herbicides are classified/grouped in various ways e.g. according to the activity, timing of application, method of application, mechanism of action, chemical family. This gives rise to a considerable level of terminology related to herbicides and their use.

Intended Outcome

- Control is the destruction of unwanted weeds, or the damage of them to the point where they are no longer competitive with the crop.
- Suppression is incomplete control still providing some economic benefit, such as reduced competition with the crop.
- Crop safety, for selective herbicides, is the relative absence of damage or stress to the crop. Most selective herbicides cause some visible stress to crop plants.
- Defoliant, similar to herbicides, but designed to remove foliage (leaves) rather than kill the plant.

Selectivity (All Plants or Specific Plants)

- Selective herbicides: They control or suppress certain plants without affecting the growth of other plants species. Selectivity may be due to translocation, differential absorption, physical (morphological) or physiological differences between plant species. 2,4-D, mecoprop, dicamba control many broadleaf weeds but remain ineffective against turfgrasses.
- Non-selective herbicides: These herbicides are not specific in acting against certain plant species and control all plant material with which they come into contact. They are used to clear industrial sites, waste ground, railways and railway embankments. Paraquat, glufosinate, glyphosate are non-selective herbicides.

Timing of Application

- Preplant: Preplant herbicides are nonselective herbicides applied to soil before planting. Some preplant herbicides may be mechanically incorporated into the soil. The objective for incorporation is to prevent dissipation through photodecomposition and/or volatility. The herbicides kill weeds as they grow through the herbicide treated zone. Volatile herbicides have to be incorporated into the soil before planting the pasture. Agricultural crops grown in soil treated with a preplant herbicide include tomatoes, corn, soybeans and strawberries. Soil fumigants like metam-sodium and dazomet are in use as preplant herbicides.
- Preemergence: Preemergence herbicides are applied before the weed seedlings emerge through the soil surface. Herbicides do not prevent weeds from germinating but they kill weeds as they grow through the herbicide treated zone by affecting the cell division in the emerging seedling. Dithopyr and pendimethalin are preemergence herbicides. Weeds that have already emerged before application or activation are not affected by pre-herbicides as their primary growing point escapes the treatment.
- Postemergence: These herbicides are applied after weed seedlings have emerged through the soil surface. They can be foliar or root absorbed, selective or nonselective, contact or systemic. Application of these herbicides is avoided during rain because the problem of being washed off to the soil makes it ineffective. 2,4-D is a selective, systemic, foliar absorbed postemergence herbicide.

Method of Application

- Soil applied: Herbicides applied to the soil are usually taken up by the root or shoot of the emerging seedlings and are used as preplant or preemergence treatment. Several factors influence the effectiveness of soil-applied herbicides. Weeds absorb herbicides by both passive and active mechanism. Herbicide adsorption to soil colloids or organic matter often reduces its amount available for weed absorption. Positioning of herbicide in correct layer of soil is very important, which can be achieved mechanically and by rainfall. Herbicides on the soil surface are subjected to several processes that reduce their availability. Volatility and photolysis are two common processes that reduce the availability of herbicides. Many soil applied herbicides are absorbed through plant shoots while they are still underground leading to their death or injury. EPTC and trifluralin are soil applied herbicides.

- Foliar applied: These are applied to portion of the plant above the ground and are absorbed by exposed tissues. These are generally postemergence herbicides and can either be translocated (systemic) throughout the plant or remain at specific site (contact). External barriers of plants like cuticle, waxes, cell wall etc. affect herbicide absorption and action. Glyphosate, 2,4-D and dicamba are foliar applied herbicide.

Persistence

- Residual activity: A herbicide is described as having low residual activity if it is neutralized within a short time of application (within a few weeks or months) - typically this is due to rainfall, or by reactions in the soil. A herbicide described as having high residual activity will remains potent for a long term in the soil. For some compounds, the residual activity can leave the ground almost permanently barren.

Mechanism of Action

Herbicides are often classified according to their site of action, because as a general rule, herbicides within the same site of action class will produce similar symptoms on susceptible plants. Classification based on site of action of herbicide is comparatively better as herbicide resistance management can be handled more properly and effectively. Classification by mechanism of action (MOA) indicates the first enzyme, protein, or biochemical step affected in the plant following application.

List of Mechanisms Found in Modern Herbicides

- ACCase inhibitors compounds kill grasses. Acetyl coenzyme A carboxylase (ACCase) is part of the first step of lipid synthesis. Thus, ACCase inhibitors affect cell membrane production in the meristems of the grass plant. The ACCases of grasses are sensitive to these herbicides, whereas the ACCases of dicot plants are not.

- ALS inhibitors: the acetolactate synthase (ALS) enzyme (also known as acetohydroxyacid synthase, or AHAS) is the first step in the synthesis of the branched-chain amino acids (valine, leucine, and isoleucine). These herbicides slowly starve affected plants of these amino acids, which eventually leads to inhibition of DNA synthesis. They affect grasses and dicots alike. The ALS inhibitor family includes various sulfonylureas (such as Flazasulfuron and Metsulfuron-methyl), imidazolinones, triazolopyrimidines, pyrimidinyl oxybenzoates, and sulfonylamino carbonyl triazolinones. The ALS biological pathway exists only in plants and not animals, thus making the ALS-inhibitors among the safest herbicides.

- EPSPS inhibitors: The enolpyruvylshikimate 3-phosphate synthase enzyme EPSPS is used in the synthesis of the amino acids tryptophan, phenylalanine and tyrosine. They affect grasses and dicots alike. Glyphosate (Roundup) is a systemic EPSPS inhibitor inactivated by soil contact.

- Synthetic auxins inaugurated the era of organic herbicides. They were discovered in the 1940s after a long study of the plant growth regulator auxin. Synthetic auxins mimic this plant hormone. They have several points of action on the cell membrane, and are effective in the control of dicot plants. 2,4-D is a synthetic auxin herbicide.

- Photosystem II inhibitors reduce electron flow from water to NADPH2+ at the photochemical step in photosynthesis. They bind to the Qb site on the D1 protein, and prevent quinone from binding to this site. Therefore, this group of compounds causes electrons to accumulate on chlorophyll molecules. As a consequence, oxidation reactions in excess of those normally tolerated by the cell occur, and the plant dies. The triazine herbicides (including atrazine) and urea derivatives (diuron) are photosystem II inhibitors.
- Photosystem I inhibitors steal electrons from the normal pathway through FeS to Fdx to NADP leading to direct discharge of electrons on oxygen. As a result, reactive oxygen species are produced and oxidation reactions in excess of those normally tolerated by the cell occur, leading to plant death. Bipyridinium herbicides (such as diquat and paraquat) inhibit the Fe-S – Fdx step of that chain, while diphenyl ether herbicides (such as nitrofen, nitrofluorfen, and acifluorfen) inhibit the Fdx – NADP step.
- HPPD inhibitors inhibit 4-Hydroxyphenylpyruvate dioxygenase, which are involved in tyrosine breakdown. Tyrosine breakdown products are used by plants to make carotenoids, which protect chlorophyll in plants from being destroyed by sunlight. If this happens, the plants turn white due to complete loss of chlorophyll, and the plants die. Mesotrione and sulcotrione are herbicides in this class; a drug, nitisinone, was discovered in the course of developing this class of herbicides.

Herbicide Group (Labeling)

One of the most important methods for preventing, delaying, or managing resistance is to reduce the reliance on a single herbicide mode of action. To do this, farmers must know the mode of action for the herbicides they intend to use, but the relatively complex nature of plant biochemistry makes this difficult to determine. Attempts were made to simplify the understanding of herbicide mode of action by developing a classification system that grouped herbicides by mode of action. Eventually the Herbicide Resistance Action Committee (HRAC) and the Weed Science Society of America (WSSA) developed a classification system. The WSSA and HRAC systems differ in the group designation. Groups in the WSSA and the HRAC systems are designated by numbers and letters, respectively. The goal for adding the “Group” classification and mode of action to the herbicide product label is to provide a simple and practical approach to deliver the information to users. This information will make it easier to develop educational material that is consistent and effective. It should increase user’s awareness of herbicide mode of action and provide more accurate recommendations for resistance management. Another goal is to make it easier for users to keep records on which herbicide mode of actions are being used on a particular field from year to year.

Chemical Family

Detailed investigations on chemical structure of the active ingredients of the registered herbicides showed that some moieties (moiety is a part of a molecule that may include either whole functional groups or parts of functional groups as substructures; a functional group has similar chemical properties whenever it occurs in different compounds) have the same mechanisms of action.

According to Forouzesh *et al.* 2015, these moieties have been assigned to the names of chemical families and active ingredients are then classified within the chemical families accordingly. Knowing about herbicide chemical family grouping could serve as a short-term strategy for managing resistance to site of action.

Use and Application

Herbicides being sprayed from the spray arms of a tractor in North Dakota.

Most herbicides are applied as water-based sprays using ground equipment. Ground equipment varies in design, but large areas can be sprayed using self-propelled sprayers equipped with long booms, of 60 to 120 feet (18 to 37 m) with spray nozzles spaced every 20–30 inches (510–760 mm) apart. Towed, handheld, and even horse-drawn sprayers are also used. On large areas, herbicides may also at times be applied aerially using helicopters or airplanes, or through irrigation systems (known as chemigation).

A further method of herbicide application developed around 2010, involves ridding the soil of its active weed seed bank rather than just killing the weed. This can successfully treat annual plants but not perennials. Researchers at the Agricultural Research Service found that the application of herbicides to fields late in the weeds' growing season greatly reduces their seed production, and therefore fewer weeds will return the following season. Because most weeds are annuals, their seeds will only survive in soil for a year or two, so this method will be able to destroy such weeds after a few years of herbicide application.

Weed-wiping may also be used, where a wick wetted with herbicide is suspended from a boom and dragged or rolled across the tops of the taller weed plants. This allows treatment of taller grassland weeds by direct contact without affecting related but desirable shorter plants in the grassland sward beneath. The method has the benefit of avoiding spray drift. In Wales, a scheme offering free weed-wiper hire was launched in 2015 in an effort to reduce the levels of MCPA in water courses.

Misuse and Misapplication

Herbicide volatilisation or spray drift may result in herbicide affecting neighboring fields or plants, particularly in windy conditions. Sometimes, the wrong field or plants may be sprayed due to error.

Use Politically, Militarily, and in Conflict

Health and Environmental Effects

Herbicides have widely variable toxicity in addition to acute toxicity from occupational exposure levels.

Some herbicides cause a range of health effects ranging from skin rashes to death. The pathway of attack can arise from intentional or unintentional direct consumption, improper application resulting in the herbicide coming into direct contact with people or wildlife, inhalation of aerial sprays, or food consumption prior to the labeled preharvest interval. Under some conditions, certain herbicides can be transported via leaching or surface runoff to contaminate groundwater or distant surface water sources. Generally, the conditions that promote herbicide transport include intense storm events (particularly shortly after application) and soils with limited capacity to adsorb or retain the herbicides. Herbicide properties that increase likelihood of transport include persistence (resistance to degradation) and high water solubility.

Phenoxy herbicides are often contaminated with dioxins such as TCDD; research has suggested such contamination results in a small rise in cancer risk after occupational exposure to these herbicides. Triazine exposure has been implicated in a likely relationship to increased risk of breast cancer, although a causal relationship remains unclear.

Herbicide manufacturers have at times made false or misleading claims about the safety of their products. Chemical manufacturer Monsanto Company agreed to change its advertising after pressure from New York attorney general Dennis Vacco; Vacco complained about misleading claims that its spray-on glyphosate-based herbicides, including Roundup, were safer than table salt and "practically non-toxic" to mammals, birds, and fish (though proof that this was ever said is hard to find). Roundup is toxic and has resulted in death after being ingested in quantities ranging from 85 to 200 ml, although it has also been ingested in quantities as large as 500 ml with only mild or moderate symptoms. The manufacturer of Tordon 101 (Dow AgroSciences, owned by the Dow Chemical Company) has claimed Tordon 101 has no effects on animals and insects, in spite of evidence of strong carcinogenic activity of the active ingredient Picloram in studies on rats.

The risk of Parkinson's disease has been shown to increase with occupational exposure to herbicides and pesticides. The herbicide paraquat is suspected to be one such factor.

All commercially sold, organic and nonorganic herbicides must be extensively tested prior to approval for sale and labeling by the Environmental Protection Agency. However, because of the large number of herbicides in use, concern regarding health effects is significant. In addition to health effects caused by herbicides themselves, commercial herbicide mixtures often contain other chemicals, including inactive ingredients, which have negative impacts on human health.

Ecological Effects

Commercial herbicide use generally has negative impacts on bird populations, although the impacts are highly variable and often require field studies to predict accurately. Laboratory studies have at times overestimated negative impacts on birds due to toxicity, predicting serious problems that were not observed in the field. Most observed effects are due not to toxicity, but to habitat changes and the decreases in abundance of species on which birds rely for food or shelter. Herbicide use in silviculture, used to favor certain types of growth following clearcutting, can cause significant drops in bird populations. Even when herbicides which have low toxicity to birds are used, they decrease the abundance of many types of vegetation on which the birds rely. Herbicide use in agriculture in Britain has been linked to a decline in seed-eating bird species which rely on the weeds killed by the herbicides. Heavy use of herbicides in neotropical agricultural areas has been one of many factors implicated in limiting the usefulness of such agricultural land for wintering migratory birds.

Frog populations may be affected negatively by the use of herbicides as well. While some studies have shown that atrazine may be a teratogen, causing demasculinization in male frogs, the U.S. Environmental Protection Agency (EPA) and its independent Scientific Advisory Panel (SAP) examined all available studies on this topic and concluded that "atrazine does not adversely affect amphibian gonadal development based on a review of laboratory and field studies."

Scientific Uncertainty of Full Extent of Herbicide Effects

The health and environmental effects of many herbicides is unknown, and even the scientific community often disagrees on the risk. For example, a 1995 panel of 13 scientists reviewing studies on the carcinogenicity of 2,4-D had divided opinions on the likelihood 2,4-D causes cancer in humans. As of 1992, studies on phenoxy herbicides were too few to accurately assess the risk of many types of cancer from these herbicides, even though evidence was stronger that exposure to these herbicides is associated with increased risk of soft tissue sarcoma and non-Hodgkin lymphoma. Furthermore, there is some suggestion that herbicides can play a role in sex reversal of certain organisms that experience temperature-dependent sex determination, which could theoretically alter sex ratios.

Resistance

Weed resistance to herbicides has become a major concern in crop production worldwide. Resistance to herbicides is often attributed to lack of rotational programmes of herbicides and to continuous applications of herbicides with the same sites of action. Thus, a true understanding of the sites of action of herbicides is essential for strategic planning of herbicide-based weed control.

Plants have developed resistance to atrazine and to ALS-inhibitors, and more recently, to glyphosate herbicides. Marestail is one weed that has developed glyphosate resistance. Glyphosate-resistant weeds are present in the vast majority of soybean, cotton and corn farms in some U.S. states. Weeds that can resist multiple other herbicides are spreading. Few new herbicides are near commercialization, and none with a molecular mode of action for which there is no resistance. Because most herbicides could not kill all weeds, farmers rotated crops and herbicides to stop resistant weeds. During its initial years, glyphosate was not subject to resistance and allowed farmers to reduce the use of rotation.

A family of weeds that includes waterhemp (Amaranthus rudis) is the largest concern. A 2008-9 survey of 144 populations of waterhemp in 41 Missouri counties revealed glyphosate resistance in 69%. Weeds from some 500 sites throughout Iowa in 2011 and 2012 revealed glyphosate resistance in approximately 64% of waterhemp samples. The use of other killers to target "residual" weeds has become common, and may be sufficient to have stopped the spread of resistance From 2005 through 2010 researchers discovered 13 different weed species that had developed resistance to glyphosate. But since then only two more have been discovered. Weeds resistant to multiple herbicides with completely different biological action modes are on the rise. In Missouri, 43% of samples were resistant to two different herbicides; 6% resisted three; and 0.5% resisted four. In Iowa 89% of waterhemp samples resist two or more herbicides, 25% resist three, and 10% resist five.

For southern cotton, herbicide costs has climbed from between \$50 and \$75 per hectare a few years ago to about \$370 per hectare in 2013. Resistance is contributing to a massive shift away from growing cotton; over the past few years, the area planted with cotton has declined by 70% in Arkansas and by 60% in Tennessee. For soybeans in Illinois, costs have risen from about \$25 to \$160 per hectare.

Dow, Bayer CropScience, Syngenta and Monsanto are all developing seed varieties resistant to herbicides other than glyphosate, which will make it easier for farmers to use alternative weed killers. Even though weeds have already evolved some resistance to those herbicides, Powles says the new seed-and-herbicide combos should work well if used with proper rotation.

Biochemistry of Resistance

Resistance to herbicides can be based on one of the following biochemical mechanisms:

- Target-site resistance: This is due to a reduced (or even lost) ability of the herbicide to bind to its target protein. The effect usually relates to an enzyme with a crucial function in a metabolic pathway, or to a component of an electron-transport system. Target-site resistance may also be caused by an overexpression of the target enzyme (via gene amplification or changes in a gene promoter).
- Non-target-site resistance: This is caused by mechanisms that reduce the amount of herbicidal active compound reaching the target site. One important mechanism is an enhanced metabolic detoxification of the herbicide in the weed, which leads to insufficient amounts of the active substance reaching the target site. A reduced uptake and translocation, or sequestration of the herbicide, may also result in an insufficient herbicide transport to the target site.
- Cross-resistance: In this case, a single resistance mechanism causes resistance to several herbicides. The term target-site cross-resistance is used when the herbicides bind to the same target site, whereas non-target-site cross-resistance is due to a single non-target-site mechanism (e.g., enhanced metabolic detoxification) that entails resistance across herbicides with different sites of action.
- Multiple resistance: In this situation, two or more resistance mechanisms are present within individual plants, or within a plant population.

Resistance Management

Worldwide experience has been that farmers tend to do little to prevent herbicide resistance developing, and only take action when it is a problem on their own farm or neighbor's. Careful observation is important so that any reduction in herbicide efficacy can be detected. This may indicate evolving resistance. It is vital that resistance is detected at an early stage as if it becomes an acute, whole-farm problem, options are more limited and greater expense is almost inevitable. Table 1 lists factors which enable the risk of resistance to be assessed. An essential pre-requisite for confirmation of resistance is a good diagnostic test. Ideally this should be rapid, accurate, cheap and accessible. Many diagnostic tests have been developed, including glasshouse pot assays, petri dish assays and chlorophyll fluorescence. A key component of such tests is that the response of the suspect population to a herbicide can be compared with that of known susceptible and resistant standards under controlled conditions. Most cases of herbicide resistance are a consequence of the repeated use of herbicides, often in association with crop monoculture and reduced cultivation practices. It is necessary, therefore, to modify these practices in order to prevent or delay the onset of resistance or to control existing resistant populations. A key objective should be the reduction in selection pressure. An integrated weed management (IWM) approach is required, in which as many tactics as possible are used to combat weeds. In this way, less reliance is placed on herbicides and so selection pressure should be reduced.

Optimising herbicide input to the economic threshold level should avoid the unnecessary use of herbicides and reduce selection pressure. Herbicides should be used to their greatest potential by ensuring that the timing, dose, application method, soil and climatic conditions are optimal for good activity. In the UK, partially resistant grass weeds such as *Alopecurus myosuroides* (blackgrass) and *Avena* spp. (wild oat) can often be controlled adequately when herbicides are applied at the 2-3 leaf stage, whereas later applications at the 2-3 tiller stage can fail badly. Patch spraying, or applying herbicide to only the badly infested areas of fields, is another means of reducing total herbicide use.

Table: Agronomic factors influencing the risk of herbicide resistance development.

Factor	Low risk	High risk
Cropping system	Good rotation	Crop monoculture
Cultivation system	Annual ploughing	Continuous minimum tillage
Weed control	Cultural only	Herbicide only
Herbicide use	Many modes of action	Single modes of action
Control in previous years	Excellent	Poor
Weed infestation	Low	High
Resistance in vicinity	Unknown	Common

Approaches to Treating Resistant Weeds

Alternative Herbicides

When resistance is first suspected or confirmed, the efficacy of alternatives is likely to be the first consideration. The use of alternative herbicides which remain effective on resistant populations can be a successful strategy, at least in the short term. The effectiveness of alternative herbicides will be highly dependent on the extent of cross-resistance. If there is resistance to a single group of herbicides, then the use of herbicides from other groups may provide a simple and effective solution, at least in the short term. For example, many triazine-resistant weeds have been readily controlled by the use of alternative herbicides such as dicamba or glyphosate. If resistance extends to more than one herbicide group, then choices are more limited. It should not be assumed that resistance will automatically extend to all herbicides with the same mode of action, although it is wise to assume this until proved otherwise. In many weeds the degree of cross-resistance between the five groups of ALS inhibitors varies considerably. Much will depend on the resistance mechanisms present, and it should not be assumed that these will necessarily be the same in different populations of the same species. These differences are due, at least in part, to the existence of different mutations conferring target site resistance. Consequently, selection for different mutations may result in different patterns of cross-resistance. Enhanced metabolism can affect even closely related herbicides to differing degrees. For example, populations of *Alopecurus myosuroides* (blackgrass) with an enhanced metabolism mechanism show resistance to pendimethalin but not to trifluralin, despite both being dinitroanilines. This is due to differences in the vulnerability of these two herbicides to oxidative metabolism. Consequently, care is needed when trying to predict the efficacy of alternative herbicides.

Mixtures and Sequences

The use of two or more herbicides which have differing modes of action can reduce the selection for resistant genotypes. Ideally, each component in a mixture should:

- Be active at different target sites.
- Have a high level of efficacy.
- Be detoxified by different biochemical pathways.
- Have similar persistence in the soil (if it is a residual herbicide).
- Exert negative cross-resistance.
- Synergise the activity of the other component.

No mixture is likely to have all these attributes, but the first two listed are the most important. There is a risk that mixtures will select for resistance to both components in the longer term. One practical advantage of sequences of two herbicides compared with mixtures is that a better appraisal of the efficacy of each herbicide component is possible, provided that sufficient time elapses between each application. A disadvantage with sequences is that two separate applications have to be made and it is possible that the later application will be less effective on weeds surviving the first application. If these are resistant, then the second herbicide in the sequence may increase selection for resistant individuals by killing the susceptible plants which were damaged but not killed by the first application, but allowing the larger, less affected, resistant plants to survive. This has been cited as one reason why ALS-resistant *Stellaria media* has evolved in Scotland recently (2000), despite the regular use of a sequence incorporating mecoprop, a herbicide with a different mode of action.

Herbicide Rotations

Rotation of herbicides from different chemical groups in successive years should reduce selection for resistance. This is a key element in most resistance prevention programmes. The value of this approach depends on the extent of cross-resistance, and whether multiple resistance occurs owing to the presence of several different resistance mechanisms. A practical problem can be the lack of awareness by farmers of the different groups of herbicides that exist. In Australia a scheme has been introduced in which identifying letters are included on the product label as a means of enabling farmers to distinguish products with different modes of action.

Farming Practices and Resistance: a Case Study

Herbicide resistance became a critical problem in Australian agriculture, after many Australian sheep farmers began to exclusively grow wheat in their pastures in the 1970s. Introduced varieties of ryegrass, while good for grazing sheep, compete intensely with wheat. Ryegrasses produce so many seeds that, if left unchecked, they can completely choke a field. Herbicides provided excellent control, while reducing soil disrupting because of less need to plough. Within little more than a decade, ryegrass and other weeds began to develop resistance. In response Australian farmers changed methods. By 1983, patches of ryegrass had become immune to Hoegrass, a family of herbicides that inhibit an enzyme called acetyl coenzyme A carboxylase.

Ryegrass populations were large, and had substantial genetic diversity, because farmers had planted many varieties. Ryegrass is cross-pollinated by wind, so genes shuffle frequently. To control its distribution farmers sprayed inexpensive Hoegrass, creating selection pressure. In addition, farmers sometimes diluted the herbicide in order to save money, which allowed some plants to survive application. When resistance appeared farmers turned to a group of herbicides that block acetolactate synthase. Once again, ryegrass in Australia evolved a kind of "cross-resistance" that allowed it to rapidly break down a variety of herbicides. Four classes of herbicides become ineffective within a few years. In 2013 only two herbicide classes, called Photosystem II and long-chain fatty acid inhibitors, were effective against ryegrass.

List of Common Herbicides

Synthetic Herbicides

- 2,4-D is a broadleaf herbicide in the phenoxy group used in turf and no-till field crop production. Now, it is mainly used in a blend with other herbicides to allow lower rates of herbicides to be used; it is the most widely used herbicide in the world, and third most commonly used in the United States. It is an example of synthetic auxin (plant hormone).
- Aminopyralid is a broadleaf herbicide in the pyridine group, used to control weeds on grassland, such as docks, thistles and nettles. It is notorious for its ability to persist in compost.
- Atrazine, a triazine herbicide, is used in corn and sorghum for control of broadleaf weeds and grasses. Still used because of its low cost and because it works well on a broad spectrum of weeds common in the US corn belt, atrazine is commonly used with other herbicides to reduce the overall rate of atrazine and to lower the potential for groundwater contamination; it is a photosystem II inhibitor.
- Clopyralid is a broadleaf herbicide in the pyridine group, used mainly in turf, rangeland,

and for control of noxious thistles. Notorious for its ability to persist in compost, it is another example of synthetic auxin.

- Dicamba, a postemergent broadleaf herbicide with some soil activity, is used on turf and field corn. It is another example of a synthetic auxin.
- Glufosinate ammonium, a broad-spectrum contact herbicide, is used to control weeds after the crop emerges or for total vegetation control on land not used for cultivation.
- Fluazifop (Fuselade Forte), a post emergence, foliar absorbed, translocated grass-selective herbicide with little residual action. It is used on a very wide range of broad leaved crops for control of annual and perennial grasses.
- Fluroxypyr, a systemic, selective herbicide, is used for the control of broad-leaved weeds in small grain cereals, maize, pastures, rangeland and turf. It is a synthetic auxin. In cereal growing, fluroxypyr's key importance is control of cleavers, *Galium aparine*. Other key broadleaf weeds are also controlled.
- Glyphosate, a systemic nonselective herbicide, is used in no-till burndown and for weed control in crops genetically modified to resist its effects. It is an example of an EPSPs inhibitor.
- Imazapyr a nonselective herbicide, is used for the control of a broad range of weeds, including terrestrial annual and perennial grasses and broadleaf herbs, woody species, and riparian and emergent aquatic species.
- Imazapic, a selective herbicide for both the pre- and postemergent control of some annual and perennial grasses and some broadleaf weeds, kills plants by inhibiting the production of branched chain amino acids (valine, leucine, and isoleucine), which are necessary for protein synthesis and cell growth.
- Imazamox, an imidazolinone manufactured by BASF for postemergence application that is an acetolactate synthase (ALS) inhibitor. Sold under trade names Raptor, Beyond, and Clearcast.
- Linuron is a nonselective herbicide used in the control of grasses and broadleaf weeds. It works by inhibiting photosynthesis.
- MCPA (2-methyl-4-chlorophenoxyacetic acid) is a phenoxy herbicide selective for broadleaf plants and widely used in cereals and pasture.
- Metolachlor is a pre-emergent herbicide widely used for control of annual grasses in corn and sorghum; it has displaced some of the atrazine in these uses.
- Paraquat is a nonselective contact herbicide used for no-till burndown and in aerial destruction of marijuana and coca plantings. It is more acutely toxic to people than any other herbicide in widespread commercial use.
- Pendimethalin, a pre-emergent herbicide, is widely used to control annual grasses and some broad-leaf weeds in a wide range of crops, including corn, soybeans, wheat, cotton, many tree and vine crops, and many turfgrass species.

- Picloram, a pyridine herbicide, mainly is used to control unwanted trees in pastures and edges of fields. It is another synthetic auxin.
- Sodium chlorate *(disused/banned in some countries)*, a nonselective herbicide, is considered phytotoxic to all green plant parts. It can also kill through root absorption.
- Triclopyr, a systemic, foliar herbicide in the pyridine group, is used to control broadleaf weeds while leaving grasses and conifers unaffected.
- Several sulfonylureas, including Flazasulfuron and Metsulfuron-methyl, which act as ALS inhibitors and in some cases are taken up from the soil via the roots.

Organic Herbicides

Recently, the term "organic" has come to imply products used in organic farming. Under this definition, an organic herbicide is one that can be used in a farming enterprise that has been classified as organic. Commercially sold organic herbicides are expensive and may not be affordable for commercial farming. Depending on the application, they may be less effective than synthetic herbicides and are generally used along with cultural and mechanical weed control practices.

Homemade organic herbicides include:

- Corn gluten meal (CGM) is a natural pre-emergence weed control used in turfgrass, which reduces germination of many broadleaf and grass weeds.
- Vinegar is effective for 5–20% solutions of acetic acid, with higher concentrations most effective, but it mainly destroys surface growth, so respraying to treat regrowth is needed. Resistant plants generally succumb when weakened by respraying.
- Steam has been applied commercially, but is now considered uneconomical and inadequate. It controls surface growth but not underground growth and so respraying to treat regrowth of perennials is needed.
- Flame is considered more effective than steam, but suffers from the same difficulties.
- D-limonene (citrus oil) is a natural degreasing agent that strips the waxy skin or cuticle from weeds, causing dehydration and ultimately death.
- Saltwater or salt applied in appropriate strengths to the rootzone will kill most plants.
- Monocerin produced by certain fungi will kill certain weeds such as Johnson grass.

Of Historical Interest and Other

- 2,4,5-Trichlorophenoxyacetic acid (2,4,5-T) was a widely used broadleaf herbicide until being phased out starting in the late 1970s. While 2,4,5-T itself is of only moderate toxicity, the manufacturing process for 2,4,5-T contaminates this chemical with trace amounts of 2,3,7,8-tetrachlorodibenzo-p-dioxin (TCDD). TCDD is extremely toxic to humans. With proper temperature control during production of 2,4,5-T, TCDD levels can be held to about .005 ppm. Before the TCDD risk was well understood, early production facilities lacked proper temperature controls. Individual batches tested later were found to have as much as 60 ppm of TCDD. 2,4,5-T was withdrawn from use in the USA in 1983, at a time of heightened public sensitivity about chemical hazards in the environment. Public concern about dioxins was high, and production and use of other (non-herbicide) chemicals potentially containing TCDD contamination was also withdrawn. These included pentachlorophenol (a wood preservative) and PCBs (mainly used as stabilizing agents in transformer oil). Some feel that the 2,4,5-T withdrawal was not based on sound science. 2,4,5-T has since largely been replaced by dicamba and triclopyr.

- Agent Orange was a herbicide blend used by the British military during the Malayan Emergency and the U.S. military during the Vietnam War between January 1965 and April 1970 as a defoliant. It was a 50/50 mixture of the *n*-butyl esters of 2,4,5-T and 2,4-D. Because of TCDD contamination in the 2,4,5-T component, it has been blamed for serious illnesses in many people who were exposed to it. However, research on populations exposed to its dioxin contaminant have been inconsistent and inconclusive.
- Diesel, and other heavy oil derivatives, are known to be informally used at times, but are usually banned for this purpose.

Glyphosate

Glyphosate (*N*-(phosphonomethyl)glycine) is a broad-spectrum systemic herbicide and crop desiccant. It is an organophosphorus compound, specifically a phosphonate. It is used to kill weeds, especially annual broadleaf weeds and grasses that compete with crops. It was discovered to be an herbicide by Monsanto chemist John E. Franz in 1970. Monsanto brought it to market in 1974 under the trade name Roundup, and Monsanto's last commercially relevant United States patent expired in 2000.

Farmers quickly adopted glyphosate, especially after Monsanto introduced glyphosate-resistant Roundup Ready crops, enabling farmers to kill weeds without killing their crops. In 2007, glyphosate was the most used herbicide in the United States' agricultural sector and the second-most used in home and garden, government and industry, and commerce. By 2016 there was a 100-fold increase from the late 1970s in the frequency of applications and volumes of glyphosate-based herbicides (GBHs) applied, partly in response to the unprecedented global emergence and spread of glyphosate-resistant weeds.

Glyphosate is absorbed through foliage, and minimally through roots, and transported to growing points. It inhibits a plant enzyme involved in the synthesis of three aromatic amino acids: tyrosine, tryptophan, and phenylalanine. Therefore, it is effective only on actively growing plants and is not effective as a pre-emergence herbicide. An increasing number of crops have been genetically engineered to be tolerant of glyphosate (e.g. Roundup Ready soybean, the first Roundup Ready crop, also created by Monsanto) which allows farmers to use glyphosate as a postemergence herbicide against weeds. The development of glyphosate resistance in weed species is emerging as a costly problem. While glyphosate and formulations such as Roundup have been approved by regulatory bodies worldwide, concerns about their effects on humans and the environment persist.

Many regulatory and scholarly reviews have evaluated the relative toxicity of glyphosate as an herbicide. The German Federal Institute for Risk Assessment toxicology review in 2013 found that "the available data is contradictory and far from being convincing" with regard to correlations between exposure to glyphosate formulations and risk of various cancers, including non-Hodgkin lymphoma (NHL). A meta-analysis published in 2014 identified an increased risk of NHL in workers exposed to glyphosate formulations. In March 2015 the World Health Organization's International Agency for Research on Cancer classified glyphosate as "probably carcinogenic in humans" (category 2A) based on epidemiological studies, animal studies, and *in vitro* studies. In November, 2015, the European Food Safety Authority published an updated assessment report on glyphosate, concluding that "the substance is unlikely to be genotoxic (i.e. damaging to DNA) or to pose a carcinogenic threat to humans." Furthermore, the final report clarified that while other, probably carcinogenic, glyphosate-containing formulations may exist, studies "that look solely at the active substance glyphosate do not show this effect." In May 2016, the Joint FAO/WHO Meeting on Pesticide Residues concluded that "glyphosate is unlikely to pose a carcinogenic risk to humans from exposure through the diet", even at doses as high as 2,000 mg/kg body weight orally.

Discovery

Glyphosate was first synthesized in 1950 by Swiss chemist Henry Martin, who worked for the Swiss company Cilag. The work was never published. Stauffer Chemical patented the agent as a chemical chelator in 1964 as it binds and removes minerals such as calcium, magnesium, manganese, copper, and zinc.

Somewhat later, glyphosate was independently discovered at Monsanto in 1970. Monsanto chemists had synthesized about 100 analogs of aminomethylphosphonic acid as potential water-softening agents. Two were found to have weak herbicidal activity, and John E. Franz, a chemist at Monsanto, was asked to try to make analogs with stronger herbicidal activity. Glyphosate was the third analog he made. Franz received the National Medal of Technology in 1987 and the Perkin Medal for Applied Chemistry in 1990 for his discoveries.

In 2008, USDA-ARS scientist Stephen O. Duke and Stephen B. Powles, an Australian weed expert — described glyphosate as a "virtually ideal" herbicide. In 2010 Powles stated: "glyphosate is a one in a 100-year discovery that is as important for reliable global food production as penicillin is for battling disease."

Chemistry

Glyphosate is an aminophosphonic analogue of the natural amino acid glycine, and like all amino acids, exists in different ionic states depending on pH. Both the phosphonic acid and carboxylic acid moieties can be ionised and the amine group can be protonated and the substance exists as a series of zwitterions. Glyphosate is soluble in water to 12 g/l at room temperature. The original synthetic approach to glyphosate involved the reaction of phosphorus trichloride with formaldehyde followed by hydrolysis to yield a phosphonate. Glycine is then reacted with this phosphonate to yield glyphosate, and its name is taken as a contraction of the compounds used in this synthesis - viz. glycine and a phosphonate.

$$PCl_3 + H_2CO \rightarrow Cl_2P(=O)\text{-}CH_2Cl$$

$$Cl_2P(=O)\text{-}CH_2Cl + 2\ H_2O \rightarrow (HO)_2P(=O)\text{-}CH_2Cl + 2\ HCl$$

$$(HO)_2P(=O)\text{-}CH_2Cl + H_2N\text{-}CH_2\text{-}COOH \rightarrow (HO)_2P(=O)\text{-}CH_2\text{-}NH\text{-}CH_2\text{-}COOH + HCl$$

The main deactivation path for glyphosate is hydrolysis to aminomethylphosphonic acid.

Industrial Synthesis

Two main approaches are used to synthesize glyphosate industrially. The first is to react iminodiacetic acid with phosphorous acid and hydrochloric acid (sometimes formed *in situ* by addition of phosphorus trichloride) via a modified Mannich reaction. Oxidation then leads to the desired glyphosate product. Iminodiacetic acid is usually prepared on-site, such as by reaction of chloroacetic acid with ammonia and calcium hydroxide to produce the calcium iminodiacetate salt and then acidification of the product.

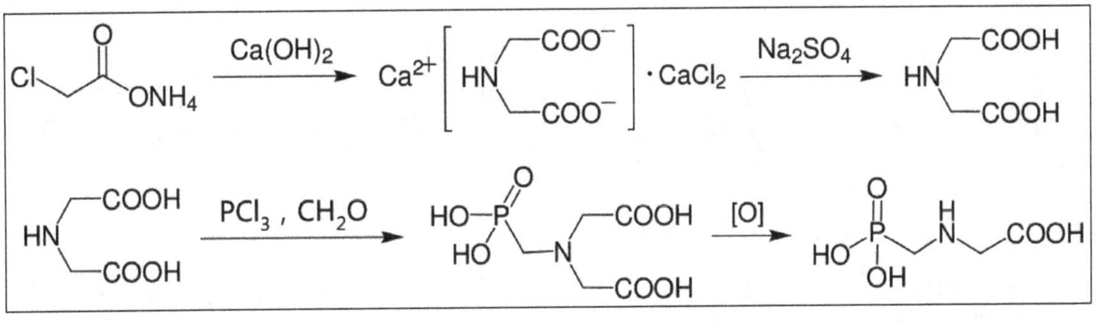

The chloroacetic acid approach is less efficient than other iminodiacetic acid approaches, owing to the production of calcium chloride waste and decreased yield. When hydrogen cyanide is readily available as a by-product (say), an alternative approach is to use iminodiacetonitrile, $HN(CH_2CN)_2$, and diethanolamine is also a suitable starting material.

The second involves the use of dimethyl phosphite in a one-pot synthesis. Glycine and paraformaldehyde are reacted in a suitable organic solvent (typically triethylamine and methanol) to produce bishydroxymethylglycine, $(HOCH_2)_2NCH_2COOH$. Dimethyl phosphite is then introduced and work-up with hydrochloric acid cleaves the hydroxymethyl group from the nitrogen atom whilst heating speeds the hydrolysis of both phosphate ester linkages.

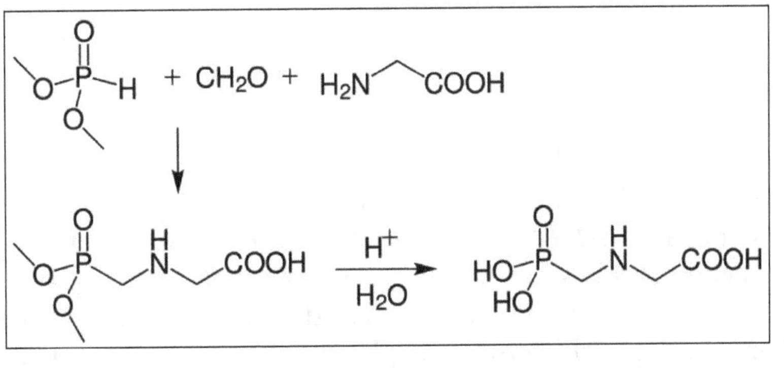

This synthetic approach is responsible for a substantial portion of the production of glyphosate in China, with considerable work having gone into recycling the triethylamine and methanol. Progress has also been made in attempting to eliminate the need for triethylamine altogether.

Mode of Action

Glyphosate kills plants by interfering with the synthesis of the aromatic amino acids phenylalanine, tyrosine, and tryptophan. It does this by inhibiting the enzyme 5-enolpyruvylshikimate-3-phosphate synthase (*EPSPS*), which catalyzes the reaction of shikimate-3-phosphate (S3P) and phosphoenolpyruvate to form 5-enolpyruvyl-shikimate-3-phosphate (EPSP). Glyphosate is absorbed through foliage and minimally through roots, meaning that it is only effective on actively growing plants and cannot prevent seeds from germinating. After application, glyphosate is readily transported around the plant to growing roots and leaves and this systemic activity is important for its effectiveness. Inhibiting the enzyme causes shikimate to accumulate in plant tissues and diverts energy and resources away from other processes. While growth stops within hours of application, it takes several days for the leaves to begin turning yellow.

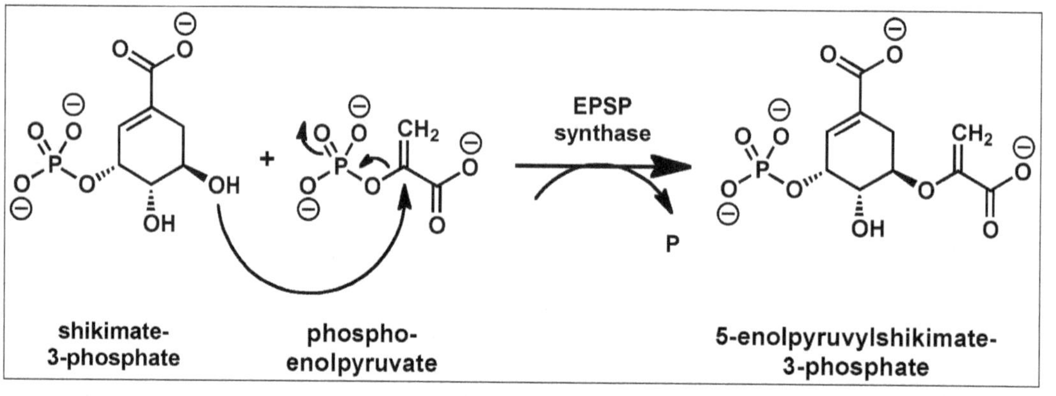

EPSP is subsequently dephosphorylated to chorismate, an essential precursor for the amino acids mentioned above. These amino acids are used in protein synthesis and to produce secondary metabolites such as folates, ubiquinones, and naphthoquinone.

X-ray crystallographic studies of glyphosate and EPSPS show that glyphosate functions by occupying the binding site of the phosphoenolpyruvate, mimicking an intermediate state of the ternary enzyme substrates complex. Glyphosate inhibits the EPSPS enzymes of different species of plants and microbes at different rates.

EPSPS is produced only by plants and microbes; the gene coding for it is not in the mammalian genome.

Environmental Fate

Glyphosate adsorbs strongly to soil, and residues are expected to generally be immobile in soil. Ground and surface water pollution is limited. Glyphosate is readily degraded by soil microbes to aminomethylphosphonic acid (AMPA, which like glyphosate strongly adsorbs to soil solids and is thus unlikely to leach to groundwater). Though both glyphosate and AMPA are commonly detected in water bodies, a portion of the AMPA detected may actually be the result of degradation of detergents rather than from glyphosate. Glyphosate does have the potential to contaminate surface waters due to its aquatic use patterns and through erosion, as it adsorbs to soil particles suspended in runoff. The mechanism of glyphosate sorption to soil is similar to that of phosphate fertilizers, the presence of which can reduce glyphosate sorption. Phosphate fertilizers are subject to release from sediments into water bodies under anaerobic conditions, and similar release can also occur with glyphosate, though significant impact of glyphosate release from sediments has not been established. Limited leaching can occur after high rainfall after application. If glyphosate reaches surface water, it is not broken down readily by water or sunlight.

The half-life of glyphosate in soil ranges between 2 and 197 days; a typical field half-life of 47 days has been suggested. Soil and climate conditions affect glyphosate's persistence in soil. The median half-life of glyphosate in water varies from a few to 91 days. At a site in Texas, half-life was as little as three days. A site in Iowa had a half-life of 141 days. The glyphosate metabolite AMPA has been found in Swedish forest soils up to two years after a glyphosate application. In this case, the persistence of AMPA was attributed to the soil being frozen for most of the year. Glyphosate adsorption to soil, and later release from soil, varies depending on the kind of soil. Glyphosate is generally less persistent in water than in soil, with 12- to 60-day persistence observed in Canadian ponds, although persistence of over a year has been recorded in the sediments of American ponds. The half-life of glyphosate in water is between 12 days and 10 weeks.

According to the National Pesticide Information Center fact sheet, glyphosate is not included in compounds tested for by the Food and Drug Administration's Pesticide Residue Monitoring Program, nor in the United States Department of Agriculture's Pesticide Data Program. However, a field test showed that lettuce, carrots, and barley contained glyphosate residues up to one year after the soil was treated with 3.71 lb of glyphosate per acre (4.15 kg per hectare). The U.S. has determined the acceptable daily intake of glyphosate at 1.75 milligrams per kilogram of bodyweight per day (mg/kg/bw/day) while the European Union has set it at 0.3.

Use

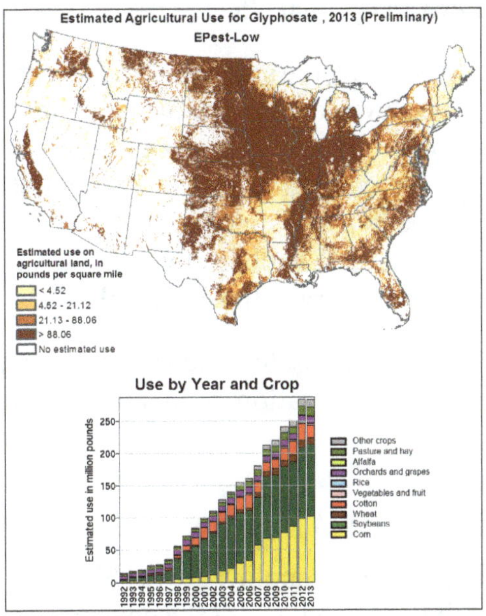

Estimated use of glyphosate in the USA in 2013 and estimated total use from 1992–2013.

Glyphosate is effective in killing a wide variety of plants, including grasses and broadleaf and woody plants. By volume, it is one of the most widely used herbicides. In 2007, glyphosate was the most used herbicide in the United States agricultural sector, with 180 to 185 million pounds

(82,000 to 84,000 tonnes) applied, the second-most used in home and garden with 5 to 8 million pounds (2,300 to 3,600 tonnes) and government applied 13 to 15 million pounds (5,900 to 6,800 tonnes) in industry and commerce. It is commonly used for agriculture, horticulture, viticulture, and silviculture purposes, as well as garden maintenance (including home use). It has a relatively small effect on some clover species and morning glory.

Glyphosate used as an alternative to mowing in an apple orchard in Ciardes, Italy.

Glyphosate and related herbicides are often used in invasive species eradication and habitat restoration, especially to enhance native plant establishment in prairie ecosystems. The controlled application is usually combined with a selective herbicide and traditional methods of weed eradication such as mulching to achieve an optimal effect.

In many cities, glyphosate is sprayed along the sidewalks and streets, as well as crevices in between pavement where weeds often grow. However, up to 24% of glyphosate applied to hard surfaces can be run off by water. Glyphosate contamination of surface water is attributed to urban and agricultural use. Glyphosate is used to clear railroad tracks and get rid of unwanted aquatic vegetation. Since 1994, glyphosate has been used in aerial spraying in Colombia in coca eradication programs; Colombia announced in May 2015 that by October, it would cease using glyphosate in these programs due to concerns about human toxicity of the chemical.

In addition to its use as an herbicide, glyphosate is also used for crop desiccation (siccation) to increase harvest yield, and as a result of desiccation, to increase sucrose concentration in sugarcane before harvest.

Genetically Modified Crops

Some micro-organisms have a version of 5-enolpyruvoyl-shikimate-3-phosphate synthetase (EPSPS) resistant to glyphosate inhibition. A version of the enzyme that was both resistant to glyphosate and that was still efficient enough to drive adequate plant growth was identified by Monsanto scientists after much trial and error in an *Agrobacterium* strain called CP4, which was found surviving in a waste-fed column at a glyphosate production facility. This *CP4* EPSPS gene was cloned and transfected into soybeans. In 1996, genetically modified soybeans were made commercially available. Current glyphosate-resistant crops include soy, maize (corn), canola, alfalfa, sugar beets, and cotton, with wheat still under development.

In 2015, 89% of corn, 94% of soybeans, and 89% of cotton produced in the US were genetically modified to be herbicide-tolerant.

Formulations and Tradenames

Glyphosate is marketed in the United States and worldwide by many agrochemical companies, in different solution strengths and with various adjuvants, under dozens of tradenames. As of 2010, more than 750 glyphosate products were on the market. In 2012, in terms of volume about half of the total global consumption of glyphosate was for conventional crops; Asia Pacific was the largest and fastest growing market. Chinese manufacturers collectively are the world's largest producers of glyphosate and its precursors and account for about 30% of global exports. Key manufacturers include Anhui Huaxing Chemical Industry Company, BASF, Bayer CropScience, Dow AgroSciences, DuPont, Jiangsu Good Harvest-Weien Agrochemical Company, Monsanto, Nantong Jiangshan Agrochemical & Chemicals Co., Nufarm Limited, SinoHarvest, Syngenta, and Zhejiang Xinan Chemical Industrial Group Company.

Glyphosate is an acid molecule, so it is formulated as a salt for packaging and handling. Various salt formulations include isopropylamine, diammonium, monoammonium, or potassium as the counterion. Some brands include more than one salt. Some companies report their product as acid equivalent (ae) of glyphosate acid, or some report it as active ingredient (ai) of glyphosate plus the salt, and others report both. To compare performance of different formulations, knowledge of how the products were formulated is needed. Given that different salts have different weights, the acid equivalent is a more accurate method of expressing and comparing concentrations.

Adjuvant loading refers to the amount of adjuvant already added to the glyphosate product. Fully loaded products contain all the necessary adjuvants, including surfactant; some contain no adjuvant system, while other products contain only a limited amount of adjuvant (minimal or partial loading) and additional surfactants must be added to the spray tank before application. As of 2000 (just before Monsanto's patent on glyphosate expired), over 400 commercial adjuvants from over 34 different companies were available for use in commercial agriculture.

Products are supplied most commonly in formulations of 120, 240, 360, 480, and 680 g/l of active ingredient. The most common formulation in agriculture is 360 g/l, either alone or with added cationic surfactants.

For 360 g/l formulations, European regulations allow applications of up to 12 l/ha for control of perennial weeds such as couch grass. More commonly, rates of 3 l/ha are practiced for control of annual weeds between crops.

Monsanto

Monsanto developed and patented the use of glyphosate to kill weeds in the early 1970s, and first brought it to market in 1974, under the Roundup brandname. While its initial patent expired in 1991, Monsanto retained exclusive rights in the United States until its patent on the isopropylamine salt expired in September 2000.

Monsanto's Roundup is the earliest formulation of glyphosate.

As of 2009, sales of these herbicide products represented about 10% of Monsanto's revenue due to competition from other producers of other glyphosate-based herbicides; their Roundup products (which include GM seeds) represented about half of Monsanto's gross margin.

The active ingredient of the Monsanto herbicides is the isopropylamine salt of glyphosate. Another important ingredient in some formulations is the surfactant polyethoxylated tallow amine.

Monsanto also produces seeds which grow into plants genetically engineered to be tolerant to glyphosate. The genes contained in these seeds are patented. Such crops allow farmers to use glyphosate as a postemergence herbicide against most broadleaf and cereal weeds. Soy was the first glyphosate-resistant crop.

References

- Bleasdale, J. K. A.; Salter, Peter John (1 January 1991). The Complete Know and Grow Vegetables. Oxford University Press. ISBN 978-0-19-286114-6.
- Ross, Merrill A.; Lembi, Carole A. (2008). Applied Weed Science: Including the Ecology and Management of Invasive Plants. Prentice Hall. p. 123. ISBN 978-0135028148.
- National Research Council (2002). Making Aquatic Weeds Useful: Some Perspectives for Developing Countries. The Minerva Group. pp. 1–9. ISBN 978-0-89499-180-6.
- National Research Council (2002). Making Aquatic Weeds Useful: Some Perspectives for Developing Countries. The Minerva Group. pp. 67–68. ISBN 978-0-89499-180-6.
- R. Goyal, Megh (2012). Management of drip/trickle or micro irrigation. Oakville, CA: Apple Academic Press. p. 104. ISBN 978-1926895123.
- Alfred J. Turgeon; Lambert Blanchard McCarty; Nick Edward Christians (2009). Weed control in turf and ornamentals. Prentice Hall. p. 126. ISBN 978-0-13-159122-6.
- Mahesh K. Upadhyaya; Robert E. Blackshaw (2007). Non-chemical Weed Management: Principles, Concepts and Technology. CABI. pp. 135–. ISBN 978-1-84593-291-6.
- Stout, Ruth. Gardening Without Work. Devon-Adair Press, 1961. Reprinted by Norton Creek Press, 2011, pp. 6-7. ISBN 978-0-9819284-6-3

Permissions

All chapters in this book are published with permission under the Creative Commons Attribution Share Alike License or equivalent. Every chapter published in this book has been scrutinized by our experts. Their significance has been extensively debated. The topics covered herein carry significant information for a comprehensive understanding. They may even be implemented as practical applications or may be referred to as a beginning point for further studies.

We would like to thank the editorial team for lending their expertise to make the book truly unique. They have played a crucial role in the development of this book. Without their invaluable contributions this book wouldn't have been possible. They have made vital efforts to compile up to date information on the varied aspects of this subject to make this book a valuable addition to the collection of many professionals and students.

This book was conceptualized with the vision of imparting up-to-date and integrated information in this field. To ensure the same, a matchless editorial board was set up. Every individual on the board went through rigorous rounds of assessment to prove their worth. After which they invested a large part of their time researching and compiling the most relevant data for our readers.

The editorial board has been involved in producing this book since its inception. They have spent rigorous hours researching and exploring the diverse topics which have resulted in the successful publishing of this book. They have passed on their knowledge of decades through this book. To expedite this challenging task, the publisher supported the team at every step. A small team of assistant editors was also appointed to further simplify the editing procedure and attain best results for the readers.

Apart from the editorial board, the designing team has also invested a significant amount of their time in understanding the subject and creating the most relevant covers. They scrutinized every image to scout for the most suitable representation of the subject and create an appropriate cover for the book.

The publishing team has been an ardent support to the editorial, designing and production team. Their endless efforts to recruit the best for this project, has resulted in the accomplishment of this book. They are a veteran in the field of academics and their pool of knowledge is as vast as their experience in printing. Their expertise and guidance has proved useful at every step. Their uncompromising quality standards have made this book an exceptional effort. Their encouragement from time to time has been an inspiration for everyone.

The publisher and the editorial board hope that this book will prove to be a valuable piece of knowledge for students, practitioners and scholars across the globe.

Index

Q

R

S

T

V

W

Y

www.ingramcontent.com/pod-product-compliance
Lightning Source LLC
LaVergne TN
LVHW080913210926
848485LV00004B/642